Mind, Stress, & Emotions

COMMONWEALTH PRESS

Boston • San Francisco

Mind, Stress, & Emotions

THE NEW SCIENCE OF MOOD

Gene Wallenstein, Ph.D.

COMMONWEALTH PRESS

Boston • San Francisco

Mind, Stress, and Emotions: The New Science of Mood

COMMONWEALTH PRESS

 Published by Commonwealth Press, Boston & San Francisco. Visit us on-line at www.commonwealthpress.net

Typeset in Adobe Garamond

Library of Congress Cataloging-in-Publication Data

Wallenstein, Gene V. (Gene Vincent), 1964-
Mind, stress, and emotions : the new science of mood / Gene V. Wallenstein.
p. cm.
Includes bibliographical references and index.
ISBN 0-97206073-1 -- ISBN 0-97206075-8 (pbk.)
1. Emotions -- Physiological aspects. 2. Neuropsychology
3. Stress (Psychology). 4. Behavior genetics. I. Title.
QP401.W35 2002
152.4--dc21 2002007081

Cover artwork © Stephanie Dalton Cowan – Getty Images, New York, NY
Edited by Nigel Goodwin – UPC, New York, NY

FOR MY FAMILY

Contents

Preface

The epidemiological numbers that describe mental illness rates in general and emotional disorders in particular are astonishingly high worldwide and future projections are even bleaker. Approximately 9.5 percent of the U.S. population aged 18-years and older suffer from some form of mood disorder in any given year. When applied to the 2000 U.S. Census residential population survey, this number translates to about 18.8 million people. This means about 1 out of every 11 American adults suffers from a clinically diagnosed mood disorder. Nearly twice as many women (12 percent) as men (6.6 percent) are affected each year.

In ***Mind, Stress, and Emotions,*** we explore the exciting new ways scientists are unraveling the secrets of mood. Beginning with the radical personality changes that occurred in the now famous railroad worker, Phineas Gage, to cutting edge research examining the genetic, biological, psychological, and environmental bases of emotions. ***Mind, Stress, and Emotions*** links important new discoveries in brain science, immunology, and stress research with novel treatment strategies for mood and anxiety disorders. In a lively, conversational tone, the reader is treated to a fascinating look at the modern science of mood in the 21st century.

INTRODUCTION

†

Autumn in Vermont

Phineas P. Gage

It's a cool crisp September afternoon in Cavendish, Vermont. The year is 1848, and like much of New England, the Green Mountain state is slowly stirring from its sleepy agrarian pace toward industrialization. To some this is a blessing, whilst other Yankees from way back see before them a world that is "in dire need of staying the same".

Part of this transformation resides in the construction of several railways that will soon connect the major cities of the region. The Rutland and Burlington line is the most southerly of the two major routes that will diagonally cut across Vermont. It begins in the little town of Bellows Falls, nestled in the southeastern corner of the state, where one can board the train at 6am and be in Boston in time for a late supper. To the north, the line moves steadily through ancient forests toward the college town of Burlington, positioned on beautiful Lake Champlain.

As one follows the planned route of the Rutland and Burlington line from Bellows Falls northward, the township of Duttonsville emerges within just a few miles, and we find ourselves abruptly shifting direction, moving on a westward path toward Proctorsville. It is in this tiny dot of a town that we find Phineas P. Gage, a foreman contracted to lay ties for this segment of our new railway.

Gage is a great favorite with the men in his gang of navigators or *navvies*, a term left over from the early days when many of the laborers working on the railway found prior employment in the construction of canals used for transporting goods. The men revere him because he is a diligent worker, "possessing an iron frame"[1], and is scrupulously fair in the way he treats those in his employ. He does not play favorites, but rather allots tasks and pay in an equitable manner. This exacting and decisive nature also makes Phineas a favorite with his employers who consider him "the most efficient and capable man"[1] they have.

Indeed, everyone seems to like Phineas Gage. Friends and neighbors describe him as "quiet and respectful of others", and such "temperate habits" are the hallmark of a good foreman. Bosses who fail to live up to these standards are unpopular, and within the culture of violence that persists among the rail workers at present, run the risk of being attacked and possibly killed. Indeed, by the time Gage is at work on the line, several foreman have already been fatally wounded by those in their charge in and around the Cavendish area.

Today is Wednesday, the 13th of September, and it is much like the days before. A noticeable difference, however, is the feeling of the air that has taken on an invigorating chill as autumn approaches. Within a month, the foliage surrounding this portion of New England will explode in a staggering palate of colors, greens gone to gold and deep reds.

At the moment, Phineas and his gang are laying a portion of the track that smacks snug up against the Black River. They will have to blast away large outcroppings of rock so that the line can assume a level and fairly straight flow. All fifty or so men on the line this day, know very well that blasting through granite is a complicated and dangerous task, only to be performed by the most experienced workers. As much an art form as engineering, the line and size of the blast must be chosen precisely to ensure the correct shaping of the rock. A blast that fragments the rock into very large pieces will make their removal more difficult and cause excessive labor and delays. Likewise, if the charge and depth of the blast are not positioned correctly, the work must be repeated in order to bring the shelf down to the desired level. Considering that it can take three men an entire day to hand drill a single hole of three-quarters inches diameter to a depth of twelve feet, talented blasters are sought after and practically worshipped by their men. And Phineas is one of the best.

Blasting involves several stages that must be performed carefully and in the correct sequence. After a small diameter hole is drilled, a safety fuse is knotted and placed into the hole. Then an explosive powder, usually made from a mixture of sulfur, charcoal, and a nitrate such as saltpeter, is placed over the fuse. Finally, sand or clay is poured over the powder and compacted with a "tamping iron". The purpose of tamping is to consolidate the explosive force to as small an area as possible, thereby allowing the charge to detonate into the rock with greater efficiency rather than escaping ineffectually back out the hole.

Phineas is an old hand at this and has had his tamping iron forged by a local blacksmith to his own specifications – "to please the fancy of the owner", others would later say. It is three feet and seven inches long, one and one-quarter inches in diameter at the larger end, and tapered to a sharp point of one-quarter inch diameter at the other end. It is quite hefty, in all weighing almost thirteen and a half pounds.

The gang has been at work all day along a bend in the bank of the river, and Phineas has just poured explosive powder into a shallow hole of about three feet. While waiting for his assistant to pour sand over the mixture before he tamps the charge, a commotion among the men erupts just a few feet behind him. Glancing back over his right shoulder, he turns to see that all is okay but must feel every ounce of the tamping iron's full weight in his hand, for it has been a long day and it is 4:30pm, almost quitting time. As he returns his attention to tamp the charge a shattering explosion occurs and his iron, sharp side forward is thrust up, piercing his left cheek. The projectile is moving with such extreme force, that it penetrates the base of his skull, plowing through the front portion of his brain, and exits out the top of his head. Phineas immediately falls back to the ground, while his tamping iron, covered with bits of blood and brain, has rocketed an additional one hundred yards before returning to earth. Amazingly he is still conscious and begins to speak to those stunned workers gathering around him.

Gage manages to stand awkwardly and "walks a few rods" of fifty feet or so to an oxcart, where he sits back against the foreboard and is driven the three-quarters of a mile back to his room in town. The cart lurches west past the intersection of Depot and Main Streets, and when it arrives at the tavern of Mr. Joseph Adams, owner and solicitor, Phineas walks to the back, allowing two of his men to help him down. He then gently moves a short distance up three stairs and comes to rest in a small chair on the tavern's veranda to await medical attention.

The first physician to arrive, some thirty minutes after the accident at 5pm, is Dr. Edward Williams, followed an hour later by a young Dr. John Harlow, just four years out of medical school. As he arrives, Dr. Williams is immediately struck by the seriousness of Phineas' condition. "He at that time was sitting in a chair upon the piazza of Mr. Adams's hotel, in Cavendish. When I drove up he said, 'Doctor, here is business enough for you.' I first noticed the wound upon the head before I alighted from my carriage, the pulsations of the brain being very distinct; there was also an appearance which, before I examined the head, I could not account for: the top of the head appeared somewhat like an inverted funnel; this was owing, I discovered, to the bone being fractured about the opening for a distance of about two inches in every direction. I ought to have mentioned above that the opening through the skull and integuments was not far from one and a half inch in diameter; the edges of this opening were everted, and the whole wound appeared as if some wedge-shaped body had passed from below upward."

Dr. Williams recalls that, "Mr. Gage, during the time I was examining this wound, was relating the manner in which he was injured to the bystanders; he talked so rationally and was so willing to answer questions, that I directed my inquiries to him in preference to the men who were with him at the time of the accident, and who were standing about at this time. Mr. G. then related to me some of the circumstances, as he has since done; and I can safely say that neither at that time nor any subsequent occasion, save once, did I consider him to be other than perfectly rational. I asked him where the bar entered, and he pointed to the wound on his cheek, which I had not before discovered; this was a slit running from the angle of the jaw forward about one and a half inch; it was very much stretched laterally, and was discoloured by powder and iron rust, or at least appeared so. Mr. Gage persisted in saying that the bar went through his head: an Irishman standing by said, 'Sure it was so, sir, for the bar is lying in the road below, all blood and brains'."[1]

Soon after Dr. Harlow arrives, he helps Phineas up the flight of stairs to his room and into bed. The scene is gruesome to those in attendance, yet Phineas appears fairly calm. Harlow, on the other hand, is somewhat horrified by the condition of his young friend. "You will excuse me from remarking here, that the picture presented was, to one unaccustomed to military surgery, truly terrific; but the patient bore his sufferings with the most heroic firmness. He recognized me at once, and said he hoped he was not much hurt. He seemed to be perfectly conscious, but was getting exhausted from the hemorrhage…"[1]

Through the next few weeks, Phineas endures convulsions, wildly fluctuating pulse changes, and a fungal infection that almost takes his life. His condition begins to worsen, and on September 24th, at 9am we find the following entry in Dr. Harlow's log: "Comatose, but will answer in monosyllables when aroused. Will not take nourishment unless strongly urged. Calls for nothing. Surface and extremities incline to be cool. Discharge from the wound scanty, its exit being interfered with by the fungi. The friends and attendants are in hourly expectancy of his death, and have his coffin and clothes in readiness to remove his remains immediately to his native place in New Hampshire. One of the attendants implored me not to do anything more for him, as it would prolong his sufferings – that if I would only keep away and let him alone, he would die. She said he appeared like "water on the brain." I said it is not water, but matter that is killing the man – so with a pair of curved scissors I cut off the fungi which were sprouting from the top of the brain and filling the opening, and made free application of caustic to them."

It is not until some two weeks later around October 6th, 23 days after the accident, that Phineas emerges from his semiconscious state and begins to stir. "General appearance somewhat improved; pulse 90, and regular; more wakeful; swelling on left side of face abating; calls for pants, and desires to be helped out of bed, though when lying upon his back cannot raise his head from the pillow."

Twenty years later in a report presented to the fellows of the Massachusetts Medical Society, Harlow looked back on the "improbability" of the case. Somehow, whilst by providence or constitution, Gage survived the remarkable accident under the talented young physician's care. "For what surgeon, the most skillful, with all the blandishments of his art, has the world ever known, who could presume to take one of his fellows who has so formidable a missile hurled through his brain, with a crash, and bring him, without the aid of this *vis conservatrix*, so that, on the fifty-sixth day thereafter, he would have been walking in the streets again? I can only say, in conclusion, with good old Ambrose Paré, I dressed him, God healed him."[1]

As improbable as the accident and recovery were, and as delighted in his progress as the physicians could be, Gage's friends soon began to realize that something in Phineas was amiss – "Gage is no longer Gage", several remarked. By November, we find Phineas in good health, but complaining of "a queer feeling in his head that he is not able to describe." The intuition by Gage and his friends that something might be wrong is, we will find, foreboding. A new Gage lurks within the old, and although he can think, feel, sense, and perceive as before, a deeply human part of Phineas has changed dramatically that will shape the remainder of his life.

Born again

Within six weeks of the accident, much of the temperament of the young foreman working on the Rutland and Burlington Railroad has now changed. Those things that made Phineas who he was, at least to his friends and family, seem to have been stripped away with the sheering force of the tamping iron. Rather than being described as one who "possessed a well-balanced mind, ...looked upon by those who knew him as a shrewd, smart businessman, [and] ...persistent in executing all his plans of operation", we find a new set of post-accident adjectives used to describe his character.

After recovering his physical strength, Phineas pleaded for his old job as foreman but was turned away. His contractors, who considered him the most efficient and capable man in their employ before the accident, regarded the change in his personality and behavior so severe that they could not grant him his position again. In a letter to the Massachusetts Medical Society Harlow writes, "the equilibrium or balance, so to speak, between his intellectual faculties and animal propensities, seems to have been destroyed. He is fitful, irreverent, indulging at times in the grossest profanity (which was not previously his custom), manifesting but little deference for his fellows, impatient of restraint or advice when it conflicts with his desires, at times perniciously obstinate, yet capricious and vacillating, devising many plans of future operation, which are no sooner arranged than they are abandoned in turn for others appearing more feasible. A child in his intellectual capacity and manifestations, he has the animal passions of a strong man."[1]

The description of the new Phineas as child-like was common among his friends, family, and even his physician. Gone was the man who routinely managed a large gang of navvies with competence and an efficiency that was the envy of other foremen. In his place we find a mercurial man who has no patience for plans or goals, little emotional attachment to previous friends, and seems to have lost the very ability to anticipate the future and organize his life accordingly. In a sense, one can imagine him being unable to relate to his own future or past, plucked from the historical continuity of his life.

Phineas, increasingly restless in the years immediately following the accident, set out on a tour of the major cities in New England, the famed tamping iron his constant companion, and eventually ended up as a resident attraction at the P.T. Barnum museum in New York. Four years after the accident, Gage, after moving through a series of short-lived farm jobs, left for South America. Although not much is known about his life during this period, it seems he continued to work as a farmhand and occasional stagecoach driver throughout Santiago and Valparaiso, Chile.

In 1859, Phineas' health began to deteriorate so he set sail for America, returning to his family who now resided in San Francisco after having joined the gold rush that was then sweeping the country. He tended the land and horses at a variety of farms in the Bay area, much like his earlier days, but soon began suffering violent convulsions. Despite medical intervention, the seizures continued to worsen until on May 21st at around 10pm Phineas finally succumbed, 13 years after that fateful day changed his life. He was 38 years old.

Dr. Harlow, now a seasoned physician still residing in Vermont, learned of Phineas' death five years later through a correspondence with Gage's mother. In what little records exist of this communication, we learn that Phineas' personality changes that occurred so abruptly following the accident persisted into his final days. "His mother, a most excellent lady, now seventy years of age, informs me that Phineas was accustomed to entertain his little nephews and nieces with the most fabulous recitals of his wonderful feats and hair-breadth escapes, without any foundation except his fancy."

Thus the change from the responsible, puritan-raised man of temperate habit, to the impulsive yarn spinning Yankee of questionable moral character was not only abrupt, but persistent as well.

What does such a dramatic change in Phineas' social conduct and indeed his psyche say about the character of personality? If we judge a person by their thoughts, deeds, and actions, who was the real Phineas Gage? And what exactly did the tamping iron remove from him as it coursed through his brain? Did it remove his ability to reason? By most accounts, the answer is no. Even Harlow had to admit that Phineas' intellectual capacities were still intact. Perhaps the iron robbed him of something more nebulous, for instance, a sense of social or moral character. If so, can we hold Phineas accountable for his post-accident transgressions? Did the injury affect his capacity for sound judgment in such a manner that one can say his "free will" was altered?

So many questions have been asked about the lessons we have really learned from Phineas Gage over the course of the last 150 years. He is undoubtedly the most famous neuropsychology patient, making appearances in the majority of introductory psychology textbooks. At the time of Harlow's reports to the Massachusetts Medical Society, fascinating discoveries were being made about the brain in other parts of the world as well. The famous neurologists, Paul Broca of France and Carl Wernicke of Germany, had by then determined that specific brain regions are responsible for the production and comprehension of language. And other reports were being made each month showing that trauma to select areas of the brain can alter various sensory and motor capacities. But, the good fellows of the Massachusetts Medical Society were not at all prepared to believe that damage to the prefrontal cortex, like that sustained to Phineas' brain, might disrupt such deeply human qualities as emotional behavior and judgment. Yet we will see in the ensuing chapters that these cousins to the human soul do indeed have a biological basis.

What this book is about

At its heart, this is a book about emotions - those sinuously elusive creatures that run afoot and afield within us all. They add a rich and complicated tapestry to our lives, while at the same time, afford us a reference point from which to understand ourselves and relate to others.

Once largely the providence of poets, philosophers, and psychologists, the study of emotions has become one of the most important and topical fields in the biological sciences, particularly neuroscience. This book examines new discoveries into how our genetic, biological, and environmental conditions generate and shape our moods. Along the way, we will also focus on practical clinical applications that arise from this research that provide a richer understanding of emotions as well as novel behavioral and pharmacological strategies for the treatment of mood disorders such as depression.

Just as a deeper understanding of emotions has contributed valuable information about mood disorders, the opposite is also true. A great deal of what we currently know about the biology of emotions has evolved from examining disturbances of normal mood patterns, not just in dramatic cases like that of Phineas Gage, but in every day folk like you and me. This is a beautiful example of how science often progresses by exploring the way exceptions help delineate the rule. Throughout this book, as we develop a feeling for the manner in which scientists currently view emotions and their disorders, we will consider the following relevant questions:

- How does stress alter the brain regions responsible for emotions?
- How does immune function interact with this process?
- Are emotions all in our heads/brains?
- How do biological and environmental conditions interact to produce emotions and their disorders?
- What can studying mood disorders teach us about the biology of emotions?
- How do learning, memory, and brain plasticity regulate mood?
- What exactly do genes encode?
- How do environmental factors regulate the expression of specific genes?
- How does learning modulate patterns of gene expression related to emotions?
- Do mood disorders have a viral cause?
- Is clinical depression an autoimmune disease?
- ... and many others. Let us get started.

Chapter 1

†

Mood Genes

Growing up as a reasonably curious kid with too much free time, I became absolutely convinced that I was adopted. I remember drilling my mother with question after question about my "real" parents - where do they live; what are they like; do they love chocolate as much as I do? I failed to see even the slightest resemblance to my parents in my own behavior, personality, or appearance. This was in sharp contrast to my best friend's family. Three generations living under the same roof and they all looked alike! More importantly, they all *seemed* alike in my eyes.

Their kitchen was my home away from home. Mrs. Mazzeo was constantly cooking, and we were constantly eating. They love food, I love food – maybe they're my real parents? I confronted my mother time and again with this and other, brilliantly reasoned theories about my real family, and she would smile with all the love, patience, and humor a parent can offer.

Looking back, I think what confused me the most was that while Mr. and Mrs. Mazzeo seemed very similar to each other (at least to me), my parents could not have been more different. My father, Eugene, was a very gregarious man who kept most of his emotional tone on a fairly level plane. My mother, Anne, by contrast is a somewhat private person whose emotional expression can be read a mile away. Whatever she feels, she shares whether she wants to or not. As I write this chapter and think back on those "formative years", it's easy to see in retrospect, elements that have emerged in my personality that appear appreciatively similar to those of my parents. Development has a way of focusing our attention on different things at different times. The older I get, the more clearly I realize two fairly fundamental things. First, that I am not as different from my parents as I once thought (God help me); and second, that my general emotional tone and the way I react to stressful events has been shaped by a complicated interaction between several factors that include (among other things): (1) my genetics; (2) my reaction to stressful events experienced throughout life, but particularly those experienced in youth; and (3) my perception of the way my parents handled stressful life events. This is not say that additional factors are unimportant. I simply want to start with these few that seem most salient. Let's begin with genetics.

Genetics and Emotional Tone

It's safe to say that one of the most misunderstood phenomena in all of science is the relationship between genes and behavior. Many people tend to think of personality, and particularly psychiatric disorders, as resulting from an additive combination of genetic constitution (nature) and life experiences (nurture). Studies of the cause of schizophrenia provide a case in point.

The first evidence that genes play a role in the development of schizophrenia came in the 1930s. One of the founders of psychiatric epidemiology, Franz Kallmann, noted that the incidence of schizophrenia is approximately 1 percent worldwide. Even though social, economic, and environmental factors vary widely from region to region across the globe, the incidence rate of schizophrenia does not. Kallmann further went on to show that the disease tends to run in families. The incidence of schizophrenia among parents, children, and siblings of patients with the disorder is 15 percent. However, the fact that the incidence of schizophrenia is more likely to run in families *does not imply a genetic basis for the disease.*

Clearly, not all conditions that run in families have a genetic cause. For example, pellagra was once a common disease with a high incidence rate within families. It is marked by nasty skin eruptions along with digestive and nervous system disturbances. We now know that pellagra is caused by a nutritional deficiency of niacin and protein in the diet. Families that were poor tended to have the disease most commonly.

Likewise, wealth and poverty are conditions that tend to run in families, but one would be hard-pressed to make the case for a genetic cause.

In order to distinguish between genetic and environmental contributions to a disease, researchers have turned to twin studies. One can compare the incidence rates of a disease in identical (monozygotic) twins, who share 100 percent of their genes, with fraternal (dizygotic) twins, who share 50 percent of their genes. (Thus fraternal twins are genetically equivalent to siblings.) If a disease is caused entirely by genetic factors, identical twins should have the same incidence rates, right? A fancier, scientific way of saying this is that the concordance rate (the tendency for twins to have the same illness) should be 100 percent.

If, on the other hand, the disease is caused by a genetic predisposition that must subsequently be triggered by an environmental factor, one would expect the concordance rate for the illness in identical twins to be less than 100 percent, but higher than that observed in fraternal twins. And this is essentially what has been shown for schizophrenia. The concordance rate is approximately 45 percent in identical twins, while only 15 percent in fraternal twins (and other siblings). Thus the tendency for sharing the disease increases with genetic similarity *within* a family.

So what does this really mean? Well, these numbers are usually interpreted as follows. First, since the concordance rate is less then 100 percent, it's thought that schizophrenia is not caused entirely by genetic makeup. Second, since the concordance rate for schizophrenia is considerably higher among identical twins in comparison to fraternal twins, there must be *some* genetic component that contributes to the development of the illness. And finally, many people make the assumption that a concordance rate of 45 percent in identical twins means that the cause of schizophrenia is determined about half by genetic/biological factors and about half by environmental/experiential factors.

This is the usual interpretation of the data, however, it has some problems. For instance, identical twins may tend to encounter similar environmental conditions more often than fraternal twins for any number of reasons. Perhaps they are treated more alike than fraternal twins by friends and family because of their appearance. Or perhaps because of the special bond of being identical twins, they seek out similar environmental situations. Even more problematic, perhaps certain personality traits are strongly influenced by genes, and these traits encourage self-selection into particular environmental situations that may trigger schizophrenia.

In this case, the genotype (a set of genes) in question is more related to the phenotype of the particular personality trait than to schizophrenia per se. This creates real problems in relating genotype to phenotype, and indeed in establishing what the phenotype actually is.

What is a phenotype?

One way to define all this phenotype business is by using what scientists refer to as the "bottom-up approach". That is, start with the lowest common denominator and work your way up. Thus in this example, the particular set of proteins that are encoded by a set of genes (the genotype) constitutes the phenotype. However, another way to define phenotype is in the overt expression of some feature or functional trait that may be related to a given genotype, for example hair color or handedness. This is the "top-down approach". In this case, one starts with a specific function or trait and works downward to the molecular mechanisms. So there are several ways one can define a phenotype.

It is yet more difficult to establish a phenotype for psychiatric disorders. This difficulty arises for a number of reasons. To begin with, psychiatric disorders such as schizophrenia often involve a complex array of symptoms. Some symptoms are more clearly observable (and quantifiable) than others. For instance, impairments in speech and motor organization are usually more overtly evident than delusions and hallucinations. Furthermore, not all patients share the same battery of symptoms.

One way to get around this is to define phenotype as the group of pathological behaviors common to *all* schizophrenics. This is a variant of the top-down approach in that one starts with a set of behaviors that are presumed to arise (at least in part) from some set of genes. Through linkage and pedigree studies, one then attempts to identify the genes that have particular variations (called alleles) that are common only to the population that expresses the phenotype. This generally sounds easier than it is. Many factors (such as personality traits) may co-vary with a given phenotype and have nothing to do with the illness. The genes that are directly related to the expression of the co-varying factors may then be mistakenly thought to be part of the genotype for the illness itself.

This example, then, has two important implications. First, that defining a phenotype can be somewhat arbitrary when done using a top-down approach. And second, that it can be misleading to think of the cause of a disease as being a simple *sum* of genetic plus environmental contributions. Indeed, this is inherently assumed when one thinks of the cause of schizophrenia as being directed by a mixture of 45 percent genetic variables and 55 percent from environmental factors. The example of individuals predisposed to schizophrenia who are also genetically endowed with a personality trait for self-selecting into environmental situations that may serve to trigger the illness (for example, being attracted to exceedingly stressful lifestyles), illustrates that what we call environmental or experiential factors interacts with ones genetic constitution. In other words, our genes not only influence the biological predisposition for developing an illness, but may also alter the likelihood of an individual being attracted to situations or stimuli that increase exposure to elements in the environment that *actually* cause the disease. Thus in a very real sense, genes affect our environmental circumstances, not to mention our perception of those circumstances.

Stress and mood genes, is there a connection?

Another way in which genes and environmental factors interact in a rather complicated manner can be illustrated by considering the relationship between stress and mood disorders. Stressful events often trigger depressive episodes in people suffering from recurring major depression, and there is fairly strong evidence that genes contribute to an individual's response to stress. Likewise, there is evidence for a genetic contribution to major depression *independent of exposure to stressors*. So, imagine that one set of genes is determined to contribute to stress liability and another set contributes to depression. The risk of onset of major depression would be similar in people with differing genetic liabilities for stress *if stressful events are not encountered*. However, in the presence of stressors, individuals with a high genetic liability to stress may be at a much greater risk for onset of an episode than those with a low stress liability. In this example, two sets of genes have a strong interaction in the eventual onset of a depressive episode, and the magnitude of the interaction depends on the presence or absence of an environmental factor (i.e. a stressful event).

Most psychiatric disorders including schizophrenia and major depression are thought to be polygenic, in that several genes may act together to control a single phenotype. Pedigree studies can be used in some cases to establish whether a disease is transferred by a dominant or recessive Mendelian inheritance pattern. However, this approach is most useful when one has an allelic variation of only a single gene that controls the expression of the phenotype. As you can imagine, most psychiatric illnesses, including mood disorders, do not display a genetic mode of transmission that is consistent with a monogenic (single gene) basis. Some researchers estimate that unipolar affective disorder may involve allelic variations in as many as 10-20 different genes, with perhaps combinations of 3-7 loci needed to cause the expression of the illness in an individual. This means that of the 10-20 possible genes related to the disorder, anywhere from 3 to 7 specific allelic variations must occur in combination to predispose an individual to major depression. Thus it is critically important to identify the genes related to a disorder; understand how they combine together to contribute to the development of the phenotype; and determine how environmental factors influence their expression.

A brief aside

Genes, DNA, RNA, and all that

Within the nucleus of a cell are chromosomes, which contain the genetic material DNA (deoxyribonucleic acid). Your particular DNA sequence was passed along to you from your parents, and comprises all the information required for manufacturing the proteins used by your body. A brain cell is distinguished from other cell types, for instance a liver cell, by the particular segment of DNA that is used to make the proteins for that cell. The "reading" of DNA is known as *gene expression*, which ultimately leads to the assembly of proteins (long strings of amino acids folded into a specific 3-dimensional structure).

Protein synthesis generally occurs within the cell, but outside the nucleus. Since DNA remains inside the nucleus of the cell, a messenger is sent to read the DNA and bring the instructions outside the nucleus for coding the actual protein. This process is carried out by messenger ribonucleic acid (mRNA). The mRNA is a sequence of nucleic acids that represents the information on a particular segment of DNA responsible for coding the manufacture of a protein. This segment of DNA is referred to as a gene. The process of assembling the piece of mRNA that contains the information in the gene is called *transcription*. Once assembled, the mRNA transcript emerges from the nucleus and is transported to sites of protein synthesis within the cell. The mRNA contains directions for assembling a chain of amino acids into a particular sequence specifying a particular protein. The process of assembling proteins based on the mRNA segments is called *translation*. This cascade of events forms the basis of what is called the "central dogma" of molecular biology, and can be illustrated as follows:

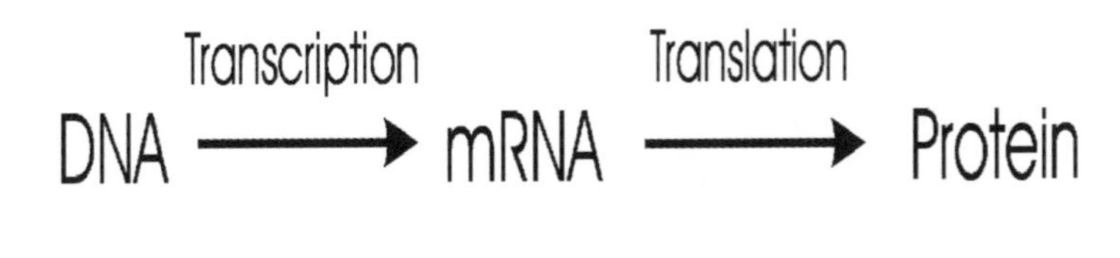

• • •

The Dual Function of Genes – what they *really* do

A great misapprehension of the role genetics plays in shaping personality stems from the idea that the sole function of genes is the transmission of hereditary information from one generation to the next. This false premise often leads to the fatalistic argument that genes are immutable to external environmental factors, and that they exert an inevitable influence on behavior. Taken to its logical extreme, this view leaves little room for social and environmental elements in shaping individual behavior, and was distorted into the eugenics movement of the 1920s and 1930s. The crucial point to remember is that genes have dual functions.

One function of genes is to operate as a template that can be reliably replicated. In this manner, the template function of genes is to pass on copies of instructions for building proteins from generation to generation. "Enlarge the nose here – yes that's good. And please widen the bald spot over there, will you?" The template is not altered by environmental experience of any kind. It is only altered through mutations, which are usually random and rare in occurrence. Thus template replication for transmission to succeeding generations is very accurate, but, of course, not perfect. Moreover, it is immutable to social forces.

The second function of genes is in determining the phenotype through what is called transcription and expression. Each cell in your body has the same complement of genes (believed to number between 50,000-200,000). The unique biological characteristics of a cell – its structure and function – are determined by the *particular set of genes* that is transcribed. Only a small fraction of the entire complement of genes within a cell is transcribed and subsequently expressed, and this gives the cell its specificity. It's what separates a brain cell from, say, a skin cell. The remaining genes that are not used in the coding of proteins for the cell are effectively repressed from being transcribed.

Now here is the interesting part. Although the template function of a gene is not modified through environmental or experiential factors, its transcription function is highly regulated by these elements. The transcription function – the ability to direct the manufacture of particular proteins through selective transcription and expression of some genes and active repression of others - is indeed *regulated by environmental factors that are experienced in a variety of different settings*. Let's take a closer look at this discovery.

How Social Factors can Influence Gene Expression

The DNA sequence of a gene can be divided up into two functionally distinct regions. The *coding region* is the portion of DNA that is transcribed by mRNA, which in turn encodes a specific protein. It is the instruction sequence or "template" for building a specific protein.

As illustrated in Figure 1-1, upstream from the coding region resides the *regulatory region*, which consists of a *promotor* and *enhancer* site. The promotor site is the location that marks the starting point for transcription of the DNA sequence, while the enhancer site recognizes special signaling proteins that determine if and when transcription occurs.

A great variety of signaling proteins can bind to specific portions of the enhancer site and thus regulate the subsequent transcription and expression of a gene. These signaling proteins are known as *transcription regulators*. Put simply, when transcription regulators bind to a segment of the enhancer site, they activate gene expression, which then results in the eventual production of a new protein. Thus they have direct control over whether a gene is turned on or off.

A number of different factors influence the way a transcription regulator binds to an enhancer site. Both internal and external stimuli (i.e. things we experience) activate signaling pathways that result in alterations of this binding process. Some signaling pathways are differentially activated as a result of the normal developmental process. Others are activated by stress, learning, hormonal changes, or social/experiential interactions.

For example, we'll see that psychological and physical stress causes the release of an adrenal gland steroid known as glucocorticoid, which circulates in the peripheral and central nervous system (brain and spinal cord). In the brain, this steroid activates a glucocorticoid transcription regulator that binds to the enhancer site of specific DNA sequences. This action induces the gene transcription and expression of new proteins involved in the long-term regulation of the stress response. This means that *social factors such as stress regulate gene expression and the subsequent production of specific proteins*.

Different stimuli, whether they originate from external environmental responses or changes within the body, that alter the binding of transcription regulators to enhancer sites of DNA affect the production of targeted proteins in all parts of the organism – literally from head to toe. They alter protein development in the body and the brain. Some enhancer sites bind transcription regulators continuously which contributes to a basal level of gene expression. Other enhancer sites bind transcription regulators in response to specific stimulation or in a periodic manner regulated by hormonal changes.

Most importantly, social influences such as learning are incorporated into our biological makeup in the altered expression of specific genes that encode the production of selective proteins in brain cells. Gene expression can be extremely selective in targeting the production of proteins unique to a specific type of nerve cell and brain region. Thus a particular experience, say a psychological stressor, will result in the production of a very specific set of proteins, while another experience, for example learning a new phone number will result in a different set.

It is important to realize that these experience-induced changes in gene expression and subsequent protein production are not transmitted from generation to generation genetically. None of these alterations in gene expression are incorporated into the sperm or egg, and therefore are not heritable. All changes in gene expression that result from learning or being exposed to experiential/environmental factors are transmitted culturally rather than genetically.

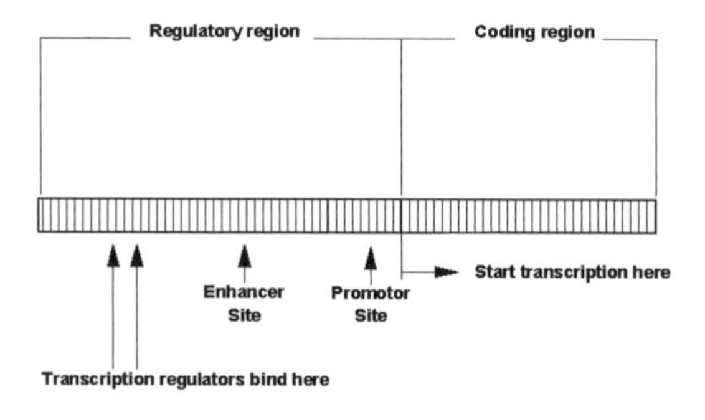

Figure 1-1 - A typical gene has two functionally distinct regions. The *coding region* contains the instructional sequence or "template" for building a specific protein. The *regulatory region* of a DNA consists of an *enhancer site*, where transcription regulators bind and determine if and when gene transcription and expression occurs. Social and environmental factors can alter the binding of transcription regulators to the enhancer site and thereby affect gene expression. These factors thus have direct control over whether a gene is expressed or repressed.

An Example: How Memories Make Proteins

The study of neural plasticity is concerned with understanding how the brain changes its physical structure in relation to normal development and in response to new experiences. One example of neural plasticity can be found in how the brain changes physically when something is learned and memorized. This is sometimes referred to as the "search for the engram". The engram being a hypothetical unit of change that occurs in the brain and represents a new memory or *memory trace*. For instance, when you learn the meaning of a new word, where is this information stored?

One of the pioneers in the hunt for the engram was the American psychologist Karl Lashley. In the 1920s, Lashley conducted a series of experiments in rats designed to test their ability to learn to find food in a maze. During their initial exposure to the task, rats take quite a bit of time to get from a starting location to a hidden food location. They routinely go down blind alleys and reach dead ends – running here and there. However, gradually, with practice rats improve their performance and learn a route that takes them straight from the starting position to the final food location.

Lashley wondered what part of their brain contains this new information. In order to find out, he trained rats to run the maze without errors and then removed a portion of their neocortex (the outer-most section of brain that is involved in a wide range of sensory-motor and higher cognitive functions). After the surgery, he placed the rat back in the starting position and recorded its behavior. Three things became apparent after several experiments.

First, rats seemed to lose their memory for the correct route following the surgery. After surgery rats routinely went down blind alleys and followed incorrect routes they had learned to avoid prior to the surgery. Second, the memory loss could be induced, by removing virtually any portion of the neocortex. Rats showed memory impairments if the lesion was in the left hemisphere, right hemisphere, temporal lobe, frontal lobe, etc. Finally, the severity of the memory loss was correlated with the size of the brain lesion (tissue destruction).

From these basic observations, Lashley came to the conclusion that a memory trace is not located in any one specific brain area, but rather is distributed equally among a large collection of regions. We now know, however, from additional studies in rats, humans, and nonhuman primates, that there do indeed seem to be specific brain areas that are required for at least some forms of memory. (Lashley may have overlooked this interpretation because most of the cortical lesions he made in his rats were very large. Thus if a smaller region that is involved in the task was destroyed by each of the larger lesions, it would appear that the memory was distributed equally to a variety of regions.)

Two key observations in the1950s suggested that indeed, some forms of memory seem to be localized and critically dependent on a region in the medial temporal lobe known as the hippocampus (Figure 1-2). Canadian neurosurgeon Wilder Penfield, working at McGill University conducted a series of cortical mapping studies in human patients undergoing brain surgery. Before excising damaged brain tissue, it's important for the surgeon to determine the functionality of the region in question. After all, it may be preferable to lose the ability to see the color green, than to lose a large portion of your visual field. This is done by delivering a brief electrical current through an electrode positioned at different locations on the cortical surface. Brain operations can sometimes be performed in alert patients under local anesthesia of the scalp, since the cortical tissue itself contains no pain receptors.

The surgeon positions the electrode, delivers a small current that will stimulate the nerve cells to fire in that region, and then asks the subject to report what they feel. Stimulating primary somatosensory regions commonly leads to patient reports of tactile stimulation on the body surface even though known is really occurring. Stimulation of motor areas often leads to muscle twitches and contractions. Stimulation of "higher" association cortical areas (so named because they receive multi-modal sensory information and are thought to serve an integrative role) often produces hallucinatory feelings that can be very complex. Some patients reported feeling a sense of déjà vu, as if they were re-living a past experience from their life. Others spoke of having flashbacks of particular recent events.

Brain scientists immediately raised the possibility that Penfield was stimulating the neural circuitry involved in storing memories for these events. Does the medial temporal lobe act like a repository for memory traces?

A man without memories

A second observation that fed into the growing excitement surrounding this issue is of a now famous patient known only by his initials, H.M, who began having seizures at the age of ten. He struggled through his adolescent years, and although medicated at near toxic levels, continued to experience seizures at greater frequency and intensity. At the age of 27, H.M. underwent a surgical procedure to remove a large portion of his medial temporal lobe from both hemispheres, which was thought to be the focal point of the seizures.

The surgery worked – his seizures began to diminish – but it soon became clear that not everything was as it should be. In virtually every way, H.M. was the same man. His intelligence, perceptual, and motor capacities remained unaltered. However, he no longer had the ability to remember new information for more than a few minutes. Brenda Milner, an experimental psychologist who was present in the operating room gallery during one of Wilder Penfield's demonstrations of stimulus-induced memory hallucinations, has studied the memory deficits in H.M. for several decades. "Removal of the medial temporal lobe didn't affect H.M.'s reasoning ability, his ability to repeat a short series of digits, or any such tasks. Incredibly, I found that he could retain a three-digit number for at least 15 minutes by continuous rehearsal, combining and recombining the digits according to an elaborate mnemonic scheme. Only when H.M. puts something out of his mind and turns to something else is the first thing lost. Say I had been working with him all morning, just the two of us, and then I'd go out for lunch. I would come back and walk by H.M. in the waiting room, and he would have no recognition of me, even though he is a very polite man. Such results appear to support the distinction between a primary memory process with a rapid decay and a secondary process (impaired in H.M.) by which the long-term storage of information is achieved."[1]

Interestingly enough, H.M. had no problems learning new procedural tasks, such as solving puzzles and performing motor actions. Indeed, he showed systematic improvement on such tasks at a rate comparable to normal subjects, despite having no recollection of having ever performed them.

Taken together, the observations of Wilder Penfield, Brenda Milner, and colleagues suggested that Lashley was probably wrong.

Certain forms of memory are indeed localized to specific brain regions such as the medial temporal lobe.

How cells learn and remember

Some 25 years later, in 1973, neuroscientists Timothy Bliss and Terje Lomo discovered a mechanism that to this day is the leading cellular model of memory. They discovered a phenomenon that is now known as long-term potentiation (LTP).

Communication between neurons is facilitated through a combination of electrical and chemical signals. When a brain cell becomes electrically excited, it produces an "action potential" that signals the release of neurotransmitter, a chemical substance such as glutamate or serotonin, which drifts across a small gap and binds to another neuron, that is referred to as the *postsynaptic cell.* We say it is postsynaptic since it is on the receiving side of the synaptic gap that separates the two neurons. The binding of transmitter to a receptor will signal changes in the postsynaptic cell that, in turn, makes it more or less excitable. Thus the communication process is an electrical signal that triggers a chemical signal that then leads to another electrical signal in the postsynaptic cell.

The change in electrical activity in the second cell due to transmitter binding to one of its receptors is known as a *postsynaptic potential.* If the postsynaptic cell becomes more excitable as a result of the transmitter binding to its receptors, the electrical change is referred to as an *excitatory postsynaptic potential (EPSP).* If, on the other hand, the postsynaptic cell becomes less excitable because of the transmitter binding to one of its receptors, the change is referred to as an *inhibitory postsynaptic potential (IPSP).*

Bliss and Lomo made the fortunate discovery that under certain conditions, the electrical potential of a postsynaptic cell is enhanced (i.e. the cell becomes even more excitable) and this "potentiation" can last days, and perhaps even longer (hence the phrase, "long-term potentiation").

This may not sound all that interesting at first glance, but the enormity of the discovery soon became apparent. It meant that theoretically a cell could undergo a long-term form of change (plasticity) that will enhance its ability to communicate with other cells following a learning experience. The question was immediately raised: is this the physical basis of memory?

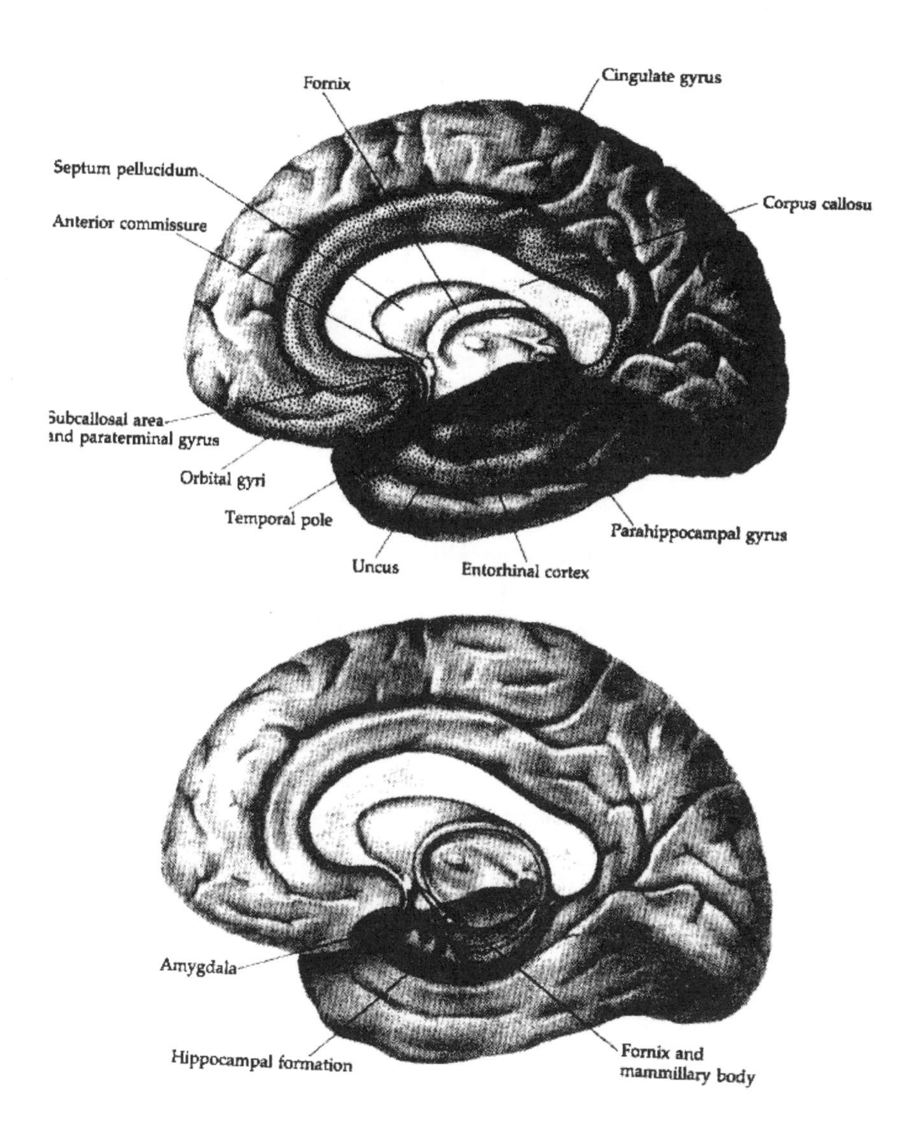

Figure 1-2 Cross-sections of the human brain showing the medial temporal lobe and associated structures. [Adapted from Martin, 1989, Human Neuroanatomy: Text and Atlas, Appleton & Lange Publishers]

Interestingly, LTP was first discovered in the hippocampus, an area of the medial temporal lobe that was already shown to be important for memory. With this finding, scientists immediately began to speculate that perhaps LTP was somehow related to long-term memory. Follow-up studies began to refine our understanding of LTP, and it now appears that for it to occur, both cells (pre and postsynaptic) must be active at the same time.

This timing constraint leads to forming a memory by association in the following way. Suppose a presynaptic cell becomes active in response to some stimulus in the environment, say the *sight* of a rose. Further, suppose a second cell (postsynaptic to the first) becomes active in response to the *smell* of a rose and another postsynaptic cell (a third cell) becomes active in response to the smell of an onion as shown in Figure 1-3.

If the subject is presented with a rose, the presynaptic cell will of course become active, as will the first postsynaptic cell (given that the subject can see and smell the rose). Moreover, assuming the rose smells like a rose and not an onion, only the first postsynaptic cell will be activated at the same time as the presynaptic cell. The fact that the presynaptic cell and the first postsynaptic cell are co-activated at the same time increases the likelihood that LTP will occur at this particular synaptic connection. Theoretically, if the enhancement of this connection is large enough, the next time the subject is presented with only the sight of a rose the activation will spread from the presynaptic cell along the potentiated pathway to re-activate the first postsynaptic cell (the cell that was activated by the smell of a rose before).

Thus after experiencing the sight and smell of a rose together, the mere visual appearance of a rose later leads to the subsequent recall of the associated smell. The smell of the onion would not be recalled with the site of the rose, because during the learning period these two phenomena did not co-occur (and the "onion cell" was not activated). This theoretical interpretation of LTP suggests that it may serve as a cellular mechanism for associative memory, and is usually captured in the phrase, "cells that fire together, wire together". Since the sight of the rose did not co-occur with the smell of the onion, the synaptic pathway connecting these two cells would not undergo LTP, and hence the normal communication between the cells would not be enhanced.

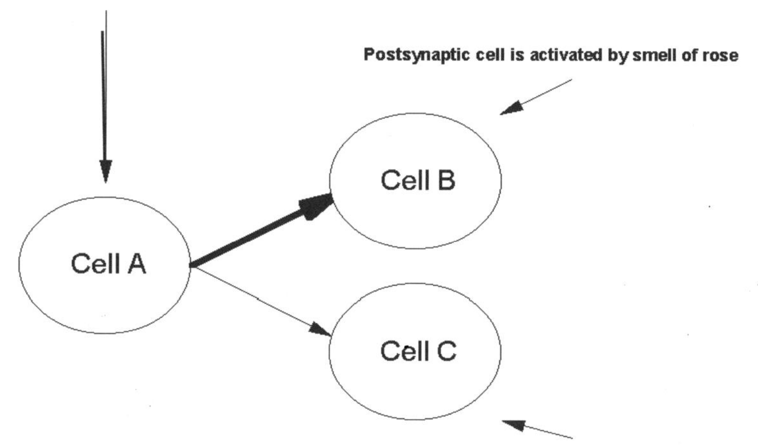

Figure 1-3 Associative memory through long-term potentiation (LTP). If a presynaptic cell (Cell A - which is activated by the *sight* of a rose) and the first postsynaptic cell (Cell B – which is activated by the *smell* of a rose) become activated at the same time due to the presence of a rose, the strength of the connection between Cells A and B increases. Once this happens, if Cell A is activated again (at a later time) by the sight of a rose, the enhanced synaptic connection with Cell B will increase the likelihood that Cell B will fire, leading to the recall (by association) of the smell of the rose. (Since the connection between Cells A and B is stronger, if Cell A becomes excited, it in turn excites Cell B.) In other words, the sight of the rose leads to a recall of its smell. In this way, LTP may form the basis of an associative memory linking the sight and smell of a rose together in a circuit that represents the "presence" of a rose. Assuming an onion was not present during the initial sighting of the rose, Cell C would not be active. Thus, the synaptic connection between Cell A and Cell C would not be enhanced. Cells that fire together, wire together. Cells that do not fire together, do not wire together.

LTP is a form of plasticity that many researchers believe represents a memory trace. In this scenario, a memory is represented by the distribution of potentiated synapses connecting cells together that were activated only by select features of a stimulus prior to learning. After learning, however, the once independent features of a stimulus (or group of stimuli) become associated with each other, and activation of any single feature (e.g. a cell that responds only to the sight of a rose) evokes activity in the entire cellular ensemble (e.g. other cells that respond to additional features associated with the presence of a rose).

While this is still a theoretical idea, there is mounting evidence, linking LTP to associative memory formation. For instance, pharmacological manipulations that block LTP from occurring also interfere with an organism's ability to form new memories. There is also data suggesting that under certain conditions, manipulations that enhance the likelihood of LTP induction also facilitate memory formation.

How memories change gene expression – flies with perfect memory

We are now beginning to understand the molecular steps involved in LTP and memory formation. That is, the physical changes that mediate this enhanced communication between cells and form the basis of a new memory. While there are many theories, the leading candidate is that new receptors must be synthesized and incorporated into the cell.

Experiments in animals support this notion to some degree. Drugs that prevent the assembly of new proteins are known as *protein synthesis inhibitors*. If these drugs are given to an animal during training on a new task, the animal learns at a normal rate, but fails to remember (compared to animals not given the drug) the task when tested days later. This suggests that protein synthesis is required for long-term memory formation.

So what does all this have to do with gene expression? What regulates the production of new proteins that is required when we learn something and have to remember it? The first step in protein synthesis is the generation of a messenger RNA transcript of a gene, and as mentioned earlier this process is controlled by transcription regulators (see Figure 1-

1). One transcription regulator that is important for memory formation is the cAMP response element binding protein (just call it "CREB"). CREB binds to the regulatory region of a gene that happens to encode proteins used in forming new memories.

There are two forms of CREB: CREB-a *activates* gene expression when it binds to the regulatory region and CREB-b *represses* gene expression. Working at Cold Spring Harbor Laboratory in New York, scientists Tim Tully and Jerry Yin performed a series of experiments designed to test the role of CREB in building memories.

They genetically engineered common fruit flies that made extra copies of the CREB-b protein and tested them to see if they could learn a simple memory task. As expected, the flies learned the procedure, but did not retain the memories as normal flies did.

Tully and Yin then genetically engineered flies that made extra copies of the CREB-a protein and ran the same tests. What do you think happened? In this case, memory tasks that normally took flies several trials to learn were remembered after a *single trial* by the mutant flies! The genetically altered flies showed perfect memory. Incredibly, these results are not restricted to flies - similar observations have now been made in marine species and mammals.

These data suggest that modulation of gene transcription and expression by CREB serves as a molecular mechanism that regulates the strength of a memory. Thus here is compelling evidence that experiential factors, such as new learning, have a direct influence on gene expression. And, to come full circle, such changes in gene expression have a direct impact on subsequent behavior (i.e. either the task is remembered or it is not, and the subject responds accordingly). Let's now consider how this relates to mood genes.

Chapter 2

†

Genes, Mood, and the Luck of the Draw

For the most part, research in psychiatric epidemiology comes in two broad categories: family and twin studies. There have been several large-scale family studies of the genetic epidemiology of major depression in the last fifteen years[1]. In a family study, one compares a proband (subjects with major depression or control subjects matched for age, gender, and other potentially confounding variables) with first-degree relatives to determine the prevalence of the disorder in biologically similar populations. While they differ on many of the quantitative estimates of hereditability of the disorder, several themes emerge consistently across studies.

First and foremost, there is consistent evidence for a strong association between mood disorder in the proband and in first-degree relatives. This supports the idea that mood disturbance runs in families. A recent review and meta-analysis of five family studies of major depression found that first-degree relatives of people suffering from major depression were approximately three times as likely to suffer from the disorder as first-degree relatives of the non-depressed comparison subjects[2].

These studies do have their limitations, however. Each of them, for example, recruited most or all of their depressed subjects from clinical referral sources. If an individual has a family history of major depression and is more likely to seek help from a clinician, this relationship will bias the analysis such that the familiality of the disorder may be overestimated. Most of the comparison subjects (who did not report having episodes of major depression) were derived from the general population rather than a medical setting. At present, it is unclear if this potential bias has had a significant influence on the overall trends in the data. At least two studies[3] did find evidence for this bias, while another did not[4].

It's also important to note that while family studies can show if a disorder runs in the family, they *do not distinguish between familial factors that are genetic from those that are due to a shared environment.* To better differentiate between these two contributing sources to the familiality of mood disorder, one needs to conduct twin and adoption studies.

Twin and adoption studies

Twin studies are designed to contrast monozygotic twins, who have essentially identical genes, with dizygotic twins, who share only half their genes. Of course, as mentioned earlier, both types of twins typically share the same environment (unless adopted), so three components of variation are important to recognize: genetic effects, environmental effects common to both members of a twin pair, and environmental effects unique to an individual member of the pair.

A recent review of six large-scale twin studies of major depression conducted in the U.S., Great Britain, Australia, and Sweden, has shown some very interesting and fairly consistent results. First, the average concordance rate (the likelihood of two members of a twin pair having the same disorder) of major depression is 45 percent in monozygotic twins and 19 percent in dizygotic twins. By gender, there is an interesting asymmetry in these data. Males have a concordance rate of 41 percent in monozygotic twins and 20 percent in dizygotic twins, while females have a concordance rate of 49 percent in monozygotic twins and 18 percent in dizygotic twins. Both groups, however, clearly show a tendency for higher concordance rates with greater genetic similarity. This implies that the etiology of major depression has a genetic component.

By considering the genetic differences between monozygotic and dyzygotic twins it is possible to partition the variation in liability to mood disorder into the three components mentioned above: genetic effects, environmental effects common to both members of a twin pair, and environmental effects unique to an individual member of the pair.

Kenneth Kendler and his colleagues working at the Virginia Institute for Psychiatry and Behavioral Genetics have performed a meta-analysis of the six major twin studies conducted in the last seven years, and extracted a summary average of these components.

Based on concordance rates and other available data, they estimate the heritability of major depression to be approximately 37 percent on average[2]. That is, approximately 37 percent of the variation in these data can be accounted for by genetic influences. (In comparison, the heritability of major bipolar disorder has been estimated to be approximately 70 percent.)

Further, they estimate that the remaining 63 percent of the variation in the data is accounted overwhelmingly by *environmental effects that are individual-specific, and not shared by both twins*. Indeed, of the six studies reviewed, only two came to the conclusion that a portion of the variance is also accounted for by environmental effects shared by both twins. The implication of this finding is that environmental influences shared by both members of a twin pair are *not* likely to have a substantial impact on their liability to major depression. These influences might include, for example, socioeconomic status, the parents' general style in regulating their children's behavior, and any number of ecological variables experienced by both twins.

It is crucial to realize that these estimates are based on an "additive model" of major depression. By that, I mean, the etiology of major depression is assumed to be due to genetic factors *plus* environmental factors. This model is based on twin studies, which are not very good at resolving gene-environment interactions. Yet, most investigators, including the authors of the above-mentioned review, recognize the importance of these possible interactions.

An example of such an interaction was given earlier, when considering the possibility that a genetic vulnerability for mood disorder may only be expressed if the individual encounters a stressful life event. Indeed, such an interaction has recently been shown by Kendler's group[5].

Gene-environment interactions that influence mood disorders

Using data from the Virginia Twin Registry, Kendler and his co-workers explored two distinct phenomena that contribute to gene-environment interactions: the "genetic control of *sensitivity* to the environment"; and the "genetic control of *exposure* to the environment".

"Genetic control of sensitivity to the environment" suggests that genes play a role in regulating the relative sensitivity of an individual to stressful life events that can trigger a depressive episode. This is contrary to the additive model, which assumes that the impact of a particular environmental variable is the same for everyone, regardless of genetic makeup.

There are certainly examples where genes alter the sensitivity of an organism to its environment. For instance, people differ substantially in the way they react to an increased dietary intake of sodium. For some, even moderate increases in sodium consumption lead to marked increases in blood pressure and hypertension. Others, however, show very little change in these variables with fairly large increases in sodium intake. This difference in the hypertensive response to sodium is controlled genetically.

We have known for quite some time that stressful life events are highly correlated with the onset of major depression. Consequently, Kendler's group assessed whether genes regulate the sensitivity of individuals to the depressogenic effects of stress. Using 2,060 monozygotic and dizygotic female twins, they split the group into four classes of genetic risk as a function of zygosity and lifetime history of major depression in the co-twin. The groups are ranked in order of increasing risk outcome as follows: (1) monozygotic twin with her co-twin unaffected; (2) dizygotic twin with her co-twin unaffected; (3) dizygotic twin with her co-twin affected; and (4) monozygotic twin with her co-twin affected. Their study, like others before, found that the onset of a major depressive episode is strongly correlated with the presence of a stressful life event. The four events with the highest correlations were: "death of a close relative", "assault", "serious marital problems", and "divorce/romantic break-up".

They then examined the risk of onset of major depression as a function of genetic risk and exposure to stressful life events. Figure 2-1 illustrates this relationship from their data. As can be seen, the risk of major depression onset varies very little with genetic liability to depression when stressful life events are absent. However, this changes dramatically when a stressor is present. In this case, the onset of a major depressive episode varies systematically with genetic liability to the disorder. In fact, with a stressor present, the increase in onset of major depression is approximately twice as high in those at greatest genetic risk as compared with those with the lowest genetic risk. This suggests a substantial interaction between a genetic (liability for major depression) and environmental (presence of a severe stressor) component in the etiology of mood disorder.

The second type gene-environment interaction that has been looked at by Kendler is the "genetic control of exposure to the environment". This hypothesis suggests that genes, at least partially, influence the onset of major depression indirectly by controlling the likelihood that an individual will be exposed to a depressogenic environment. That is, genes may regulate, to some extent, an individual's self-selection into high-risk or stressful environmental contexts.

The typical example usually given to illustrate this point is the relationship between cigarette smoking and lung cancer. There is considerable evidence suggesting that the risk for cigarette smoking is influenced, in part, by familial liability that includes genetic factors[6]. There are also a growing number of research studies listing cigarette smoking as a high risk factor in the development of lung cancer. Now imagine that you're a scientist looking for the set of genes that influences the development of this disease, but unknowingly discovered a genotype that increases the risk for cigarette smoking. Your natural tendency might be to assume that these genes directly predispose an individual to lung cancer. But this certainly would not be true in the classical way we think of the relationship between genotype and phenotype. That is the actions of the gene would not be to directly encode a protein that is involved in the biological cascade of reactions that lead to the growth of cancerous cells. Rather, the genes you have found may encode proteins that, together with many other factors, influence the likelihood that a person will be attracted to and enjoy cigarette smoking. Put simply, this set of genes might increase the probability that an individual pursues a high-risk environment for the expression of lung cancer. Because this set of genes codes for exposure to environmental factors that facilitate the risk for lung cancer, a host of additional experiential constraints becomes important.

For instance, if tobacco companies increase the price of cigarettes to offset the cost of legal settlements, one can imagine this may reduce the availability of cigarettes to poorer individuals. If poor individuals reduce their exposure to cigarette smoking they may, in turn, reduce their risk for lung cancer. Thus, in this example, an economic variable (wealth) would interact with the genes that influence the probability of exposure to cigarette smoking. This fine difference ultimately becomes very important. For the poor who discontinue smoking for economic reasons, the genes that influence exposure to cigarette smoking would not have a very large impact on the risk of developing lung cancer. However, those individuals who can afford cigarettes would be free to pursue this interest, and thus exposure to smoking would increase. In this scenario, the wealthier brethren of those genetically inclined toward smoking would have a higher risk for lung cancer. So, in this case, the genes that influence exposure to cigarette smoking may have a dramatic impact on the eventual expression of the disease.

Is it plausible that a similar phenomenon occurs in the etiology of mood disorders? Do certain individuals have a greater tendency to be exposed to stressful events? Or is the probability of experiencing a stressful life event essentially the same throughout the population? In this sense, one might think of stressors as random events and that we all get an approximately equal share of them. This is almost certainly not true. It's fairly easy to think of circumstances, perhaps economic in nature, which can increase an individual's exposure to repeated stressful events. Other examples include, living in a high crime neighborhood, being drafted into the military during wartime, or living with a chronic health condition, to name just a few.

Yet there are other circumstances that may involve a degree of self-selection into stressful contexts. For example, an individual's propensity toward choosing certain careers that may involve a high degree of stress such as being a police office or emergency room physician could clearly influence their exposure.

But what about just plain old bad luck? Interestingly enough, the presence of stressful life events is significantly correlated within an individual across a lifetime[7]. That is, a person's tendency (or lack thereof) toward encountering stressful events during one period of life is positively correlated with the tendency (or lack thereof) in other periods of life. This pattern of results is not consistent with the idea that we are passive recipients of essentially random bad luck. Indeed, some investigators have argued that the propensity toward encountering stressful life events can be predicted from relatively stable situational and personality characteristics[8].

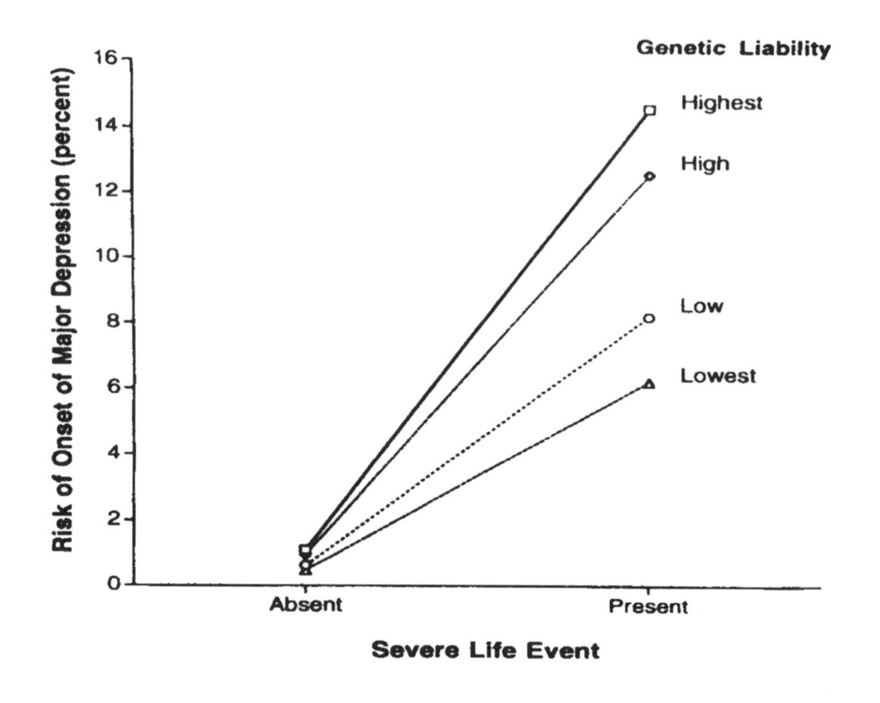

Figure 2-1 A genetic-environment interaction. The risk of major depression onset varies very little between groups with different genetic liabilities to the disease in the absence of a stressful life event. However, in the presence of a stressor, risk onset varies considerably between these groups. Indeed, the group with the highest genetic liability has a risk of onset that is approximately twice as large as the group with the lowest genetic liability. [Reproduced with permission from Kendler, 1998, *Pharmacopsychiatry*, 31, 5-9.]

Several groups have addressed this question by conducting a standard genetic analysis using twins or families, and treating the variable, "exposure to stressful life events" as any other phenotype[9]. Each of the studies performed in the last fourteen years has suggested that familial and/or genetic factors contribute to an individual's risk for encountering stressful life events[9].

Kendler's group, for example, returned to the Virginia Twin Registry database, and found that the number and magnitude of stressors encountered across a lifetime was highly correlated in twins, suggesting that exposure to stressful life events is familial. Moreover, the correlation was considerably higher in monozygotic twins (+.429) than in dizygotic twins (+.309). This implies that *similarity in rates of exposure to stressful life events increases with genetic similarity.* This is not to say that genes are the most important factor in determining exposure to stressors across a lifetime. There are clearly circumstances where a stressor affects the whole family (e.g. death of a relative), and would contribute to variation in the expression of this phenotype as an "environmental influence shared by both members of a twin pair". Indeed, in this report, the authors estimated the heritability of being exposed to stressful life events at approximately 26 percent. That is, 26 percent of the variation in this data can be accounted for by genetic variations. Approximately 18 percent of the variation was accounted for by environmental influences common to both twins, and 57 percent of the variation was accounted for by environmental factors that were specific to an individual.

So, it appears that the most significant element in determining one's risk for encountering stressful life events is indeed the idiosyncratic environmental factors that "just happen" in life. But the data also suggest a significant familial component that is not accounted for by environmental factors, either common to both twins or specific to an individual. Thus *genes do indeed influence the overall risk for encountering stressful life events.*

Considering this, what portion of the genetic variability associated with the risk of mood disorder can be attributed to an increased exposure to stressful life events? Put another way, do the genes that are associated with an increased likelihood of being exposed to a stressor overlap with those that contribute to major depression? Indeed, it does appear that the risk an individual has for experiencing certain stressful life events can be predicted by that person's level of genetic liability to major depression.

Co-twins of twins who suffer from major depression show an increased number of stressful life events compared to the normal population, and this effect is greater in monozygotic twins than in dizygotic twins. This effect is not due to stressors encountered during a depressive episode, but rather typically *precede its onset.* In particular, the genetic liability to major depression is associated with a significant increase in risk for seven stressful life events: assault, serious marital problems, divorce/loss of a significant relationship, job loss, serious illness, major financial problems, and interpersonal relationship difficulties.

In summary, the mode of inheritance for mood disorders is complex, and transmission does not seem to follow simple Mendelian patterns in most families. Linkage studies, which look for variations in specific genes, have failed to show a single, consistent genetic locus associated with major depression. There has been a report of an association between major depression and a polymorphism in the human serotonin transporter gene on chromosome 17q, but this study has not been replicated[10]. Likewise, an association has been discovered between the dopamine D4 receptor gene and major depression (which we will discuss in a later chapter), but has not been replicated[11].

The difficulties in identifying the gene action involved in major psychiatric illnesses are enormous. For mood disorders, these difficulties stem from several sources that have already been discussed. The potential multigenic locus of the disorder, together with its complicated set of symptoms makes heritability estimates highly susceptible to changes in phenotype definition. This creates a special challenge in deriving *the* genotype for mood disorder. These problems must be addressed if we are to have any real understanding of the genetic factors that contribute to the etiology of mood and its disorders.

Clearly, the findings to date also call for a greater appreciation of the subtle ways in which genetic and environmental/experiential factors interact in the genesis of mood and its disorders. A more complete account must be made of the way gene action influences our susceptibility and exposure to environmental risk factors, and the reciprocating effect the latter has on subsequent gene expression patterns. Understanding this complicated interaction is particularly informative with respect to the relationship between stress and the development of an affective disorder, which is where we turn next.

Chapter 3

†

stress, Stress, STRESS...!

It's difficult to imagine a more exciting area of biological study in the 21st century than that of mind-body interactions. This is not to say I think the mind is separable from the brain. For me, and I would guess most neuroscientists, the mind is ultimately an expression of brain mechanisms. Rather, what I wish to emphasize here is that what we think, feel, and perceive influences a number of fundamental regulatory processes of the body. These include basic immunological and endocrine functions, such as body temperature, heart rate, blood pressure, levels of assorted circulating hormones, and many others.

Likewise, the homestatic and allostatic mechanisms that regulate bodily functions can influence how we think, learn, remember, and perceive the world around us. The mind influences the body and the body influences the mind. It sounds like a cliché now, but this interaction was under appreciated at best in early biological models of the body's stress response. Enmeshed in this new understanding is a broader and richer definition of stress itself.

From an evolutionary perspective, it's easy to see that the stress response – your body's reaction to a stressor – can be a very useful thing. All organisms need a monitoring system capable of responding to environmental challenges that threaten their survival. These challenges come in many shapes and forms.

For instance, imagine you're frolicking through the park one day (perhaps in the very merry month of May), when you are taken by surprise by a pair of roguish eyes ... that scare the hell out of you. Your immediate reaction is to run for your life, and several changes in your body are needed to facilitate this course of action.

First, your body needs immediate energy if it is to escape this emergency situation. This energy comes in the way of increased glucose mobilization from your fat cells, liver and muscles to drive the body systems involved in the escape. In order to deliver the increase in glucose to your muscles as quickly as possible, arterial blood pressure, heart rate, and breathing rate increase to enhance oxygen and nutrient transport. Whilst all this is going on, other bodily functions are curtailed for the sake of energy conservation. Immune response, digestion, and sexual behavior are inhibited, along with protein synthesis involved in long-term development and growth. This makes sense, right? The last thing you're worried about when looking at those scary roguish eyes is the rate at which your Big Mac is being metabolized. And you're not exactly "in the mood for love". You need to keep your energy focused on the problem in front of you. All organisms, from simple bacteria through primates have some type of stress response system, though they may differ in their degree of complexity. Our bodies have evolved very elaborate mechanisms for effectively dealing with acute physical stressors such as the one encountered above.

A very similar set of physiological stress response actions occur when we encounter *psychological or social stressors* rather than physical stressors. With notable exceptions, our society has developed to the degree that, as humans, we no longer need to deal on a day-to-day basis with most acute physical stressors that shaped the evolution of the body's stress response. Most of us have adequate food and shelter, and rarely are we chased across the savannah by a predator. We have removed a sufficient number of the physical threats from our lives, and improved basic health conditions to the degree, that we now live long enough and have ample free time on our hands to invent a spectacularly impressive array of psychological stressors. Thus from the vantage point of human evolutionary history, psychological stressors can be considered a relatively new addition to the list of stimuli that evoke a stress response. Most animals, after all, are not overly concerned about how their 401K is doing, or whether or not they should re-mortgage their home since the federal reserve just dropped interest rates by a quarter of a percentage point.

Psychological stressors turn on essentially the same sequence of adaptive physiological responses as that of a physical threat, however, they are far more insidious in that they often lead to chronic activation of the response. This is because physical and psychological stressors operate on completely different timescales and with different energy demands. After a physical stressor appears, we get revved up with extra blood glucose and increased energy transport, and then (hopefully) use this adaptive response to escape the threatening situation. This typically happens quickly throughout the animal kingdom, and it is wonderfully adaptive in that the physiological changes that occur are required to drive the escape from the stressor. The physical demands of the escape behavior draw on these alterations in metabolic resources and once the behavior is performed, the physiological changes return to normal baseline levels.

However, this is not the case with most psychological stressors. Indeed many individuals display an exaggerated stress response, as measured by circulating blood glucocorticoids or systolic blood pressure, for very prolonged periods of time. We are talking months here.

This may be thought of as a chronic, low-level activation of the stress response system that initiates all the same endocrine and immune reactions that normally operate for a brief period of time and then turn off. As we shall see, this sustained activity of the stress response system can eventually lead to persistent changes in our body's reaction to stressful situations in the future and some have agued that these physiological adaptations may predispose certain people to mood and anxiety disorders. These long-lasting physiological changes in the body's response to stress can lead to a host of alterations in the way the body deals with disease states and infections. These are the primary concerns addressed in Robert Sapolsky's wonderful book, *Why Zebras don't get Ulcers*[1]: "You sit in your chair not moving a muscle, and simply think a thought, a thought having to do with your feeling angry or sad or euphoric or lustful, and suddenly your pancreas secretes some hormone. Your *pancreas*? How did you do that with your pancreas? You don't even know where your pancreas is. Your liver is making an enzyme that wasn't there before, your spleen is faxing a message to your thymus gland, blood flow in little capillaries in your ankles has just changed. All from thinking a thought."

The Stress Response

Our response to stressors, real or imagined, is a complex mixture of reactions that originate in a very small region, buried deep within the brain, called the *hypothalamus*. Figure 3-1 shows a brain that has been split down the middle of its long axis. The hypothalamus is buried beneath multiple layers of neocortex and located on the top portion of the brain stem (the stalk-like section that extends upward from your spinal cord).

The hypothalamus is involved in a wide variety of regulatory processes basic to our survival. All mammals live most comfortably within a specific optimal range of values for variables such as core body temperature, blood composition, pressure, and others. Brain cells, for example, are very temperature sensitive. Their electrophysiological responsiveness to stimulation changes dramatically with even small deviations away from a core body temperature of 37 degrees Celsius.

Cells in the hypothalamus are sensitive to temperature deviations and send feedback signals into the peripheral nervous system to make compensatory adjustments to the rest of your body that are designed to bring body temperature back into the optimal range. If you're a member of the Polar Bear Club and are just finishing a swim in Lake Michigan in February, upon leaving the water, your hypothalamus will scream at your autonomic nervous system to make adjustments. It will send signals whose end results are to make you shiver (in an attempt to warm the muscles), develop goose bumps (to fluff up your non-existent fur – this works best in hairier mammals), and turn your skin blue (a result of blood moving away from cold surface tissues in order to warm the inner sensitive core of the body).

If, on the other hand, you are involved in strenuous exercise on a very warm day, the hypothalamus activates systems to dissipate heat such as perspiration (which cools the skin by evaporation) and increasing blood flow to the skin surface where heat can be radiated away (and make your face appear flushed). There are many other examples of homeostatic control by the hypothalamus, including the regulation of blood oxygen, volume, salinity, pressure, acidity and so forth.

We now know that hypothalamic control of these regulatory processes is not toward a single equilibrium point in all cases. The optimal value, or "set point" can change with a number of circumstances such as time of day, whether or not you have just eaten, and so on. We call this pattern of changing set points, "allostasis" - different circumstances demand different homeostatic set points. Within this context, a stressor can be defined as anything that perturbs your body out of allostasis.

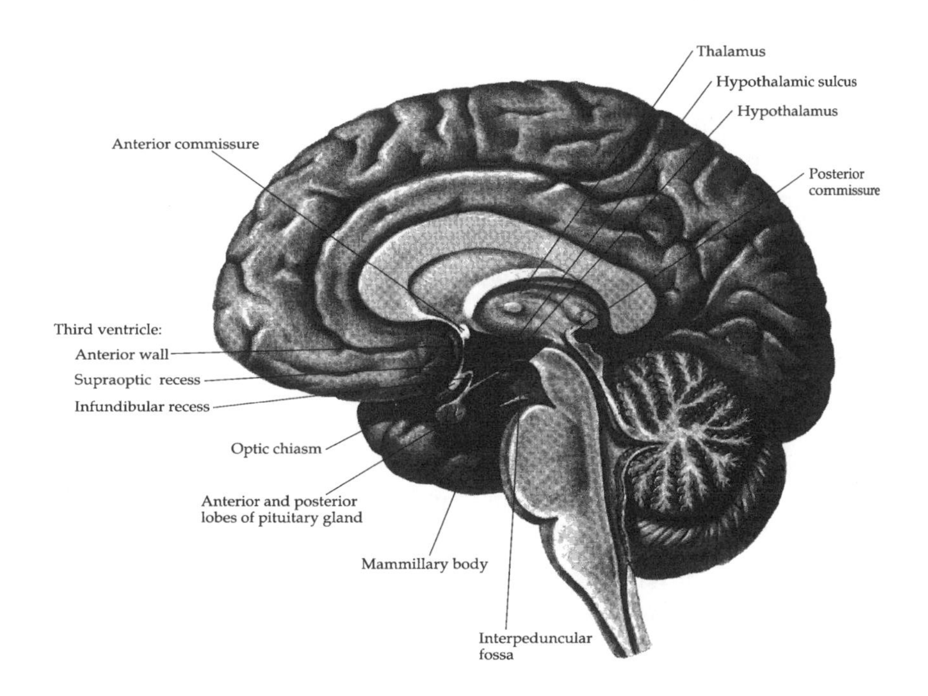

Figure 3-1 A midsagittal view (split down the long axis) of the brain, highlighting the key structures surrounding the hypothalamus and pituitary gland. The front of the brain is to the left. [Adapted from Martin, 1996, *Neuroanatomy: Text and Atlas*, Appleton and Lange Publishers]

So where do things go from the hypothalamus? The hypothalamus can be divided into three distinct zones: lateral, medial, and periventricular. I won't bore you with a lot of anatomical details, for our purposes of understanding the stress response the exciting action takes place in the periventricular region. This area is called *periventricular*, because it runs adjacent to the third ventricle ('peri' = around and 'ventricular' pertains to the brain's ventricles). Ventricles are hollow spaces in the brain that contain circulating cerebral spinal fluid, a nutrient-rich goo that contains the numerous ions used by nerve cells. (Early brain theorists, upon performing dissections of animals, discovered these fluid-filled regions and thus began the ventricular or "humoural" theory of brain function. Some theorists went on to suggest that an over-abundance of black bile causes melancholy.)

Cells in the periventricular region of the hypothalamus are in charge of two separate but related processes that are initiated when you encounter a stressor. The first concerns the hypothalamic communication with the pituitary (stay tuned), and the second process concerns the hypothalamic activation of the autonomic nervous system.

The Hypothalamus-Pituitary-Adrenal Axis

The pituitary dangles down just below the hypothalamus at the base of the brain. It is through the pituitary that the hypothalamus communicates with the rest of the body, sending hormonal messages to all points North, South, East, and West.

The pituitary has two lobes, anterior and posterior, that are controlled by the hypothalamus in different ways. The posterior lobe is considered to be part of the brain, and is involved in several regulatory functions such as controlling proper blood volume and salt concentration. The anterior lobe of the pituitary is more gland-like and is critically involved in the control of your endocrine system.

So what happens here? Cells in the periventricular region of the hypothalamus contain a hormone called corticotropin releasing factor (CRF), which they secrete into the pituitary when we encounter a stressor[2]. The cells of the pituitary are busy little guys, particularly in the perpetually stressed among us. These cells synthesize and, in turn, secrete a number of hormones that regulate many additional glands throughout the body, including the gonads, the thyroid, adrenal, and mammary glands. *The pituitary and the glands under its control form the endocrine system*[3].

Of special interest to us, is the regulation of the adrenal gland by the pituitary. The adrenal glands are located just above your kidneys and consist of two parts, a surrounding shell called the *adrenal cortex*, and a middle area called the *adrenal medulla* (hang in there, this distinction becomes important). As illustrated in Figure 3-2, when CRF binds to receptors on cells in the pituitary, they become activated and release yet another hormone called adrenocorticotropin hormone (ACTH). ACTH enters the general circulation (the bloodstream) and travels to the adrenal cortex, where within minutes, it stimulates the release of (you guessed it, yet another hormone) glucocorticoid (also known as cortisol). Cortisol travels in the bloodstream throughout the body and back into the brain, where it has a variety of effects – some good and some bad.

Cortisol is essentially a steroid, and when it's released into general circulation it has energizing effects that help prepare us to deal with stressful situations. It causes an increase in the circulating levels of blood sugar glucose, and results in a temporary suppression of the immune system, among other effects.

Cortisol levels are maintained within a specific physiological range through a negative feedback system. Once cortisol concentrations increase in general circulation, it binds to glucocorticoid receptors in the pituitary, hypothalamus, and the hippocampus, and this reduces activity in the entire stress circuit (the hypothalamus-pituitary-adrenal axis or 'HPA axis'), thereby shunting further release of cortisol. As we will see shortly, this negative feedback system may be damaged in some individuals who have endured very traumatic experiences, and is often weakened in patients with mood and anxiety disorders.

Activation of the hypothalamic-pituitary-adrenal (HPA) axis is a classical signature of the stress response, so much so that measuring circulating levels of cortisol in blood, urine, or cerebral spinal fluid is the primary means by which researchers quantify the physiological effects of a stressor[4]. So if you've skipped all the above techno-babble and have arrived at this sentence in a state of desperate confusion, the simple story is that stress turns on the HPA axis and increases levels of circulating cortisol (Figure 3-2).

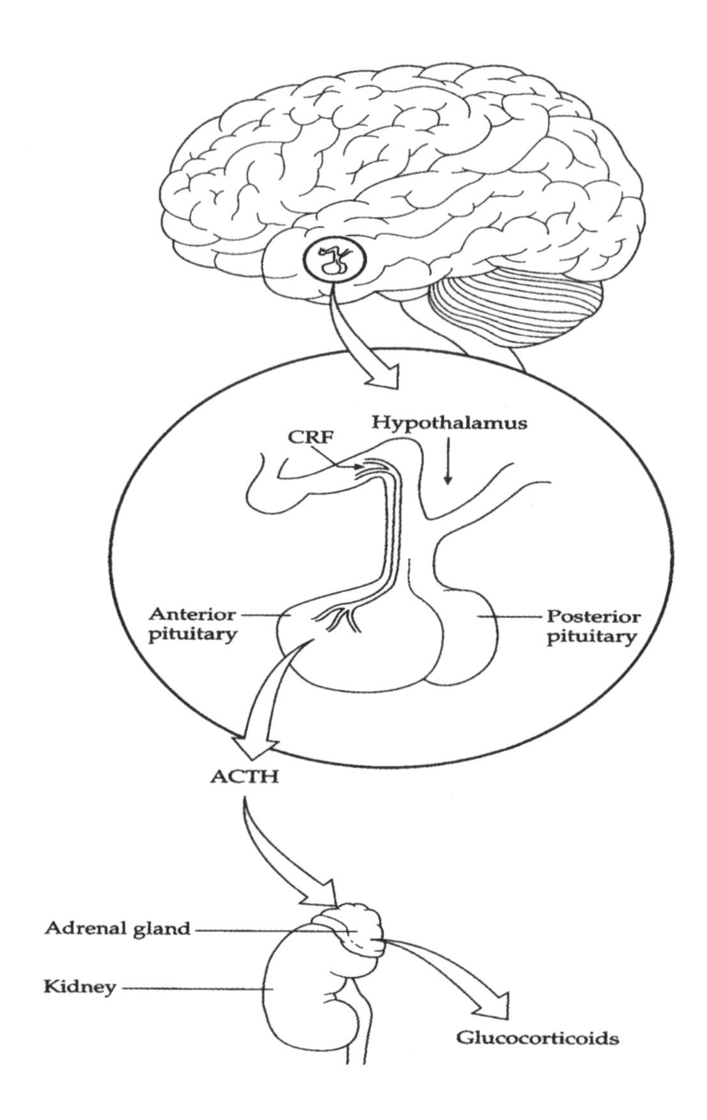

Figure 3-2 – The Hypothalamic-Pituitary-Adrenal (HPA) Axis. When a stressor is encountered, cells in the periventricular region of the hypothalamus secrete corticotropin releasing factor (CRF), a hormone, which travels to the anterior lobe of the pituitary gland through a system of small blood vessels called the hypothalamo-pituitary portal. Once CRF binds to receptors on cells

in the anterior pituitary, they, in turn, release another substance, adrenocorticotropin hormone (ACTH), that enters general circulation and eventually activates the adrenal gland located just above the kidney. The adrenal gland then releases glucocorticoids (cortisol in primates) into the bloodstream, where they travel throughout the body, producing a number of effects such as immune suppression, and increasing blood glucose. [Reproduced from Sapolsky, 1998, *Why Zebras Don't Get Ulcers*, Freeman and Company]

A fascinating aspect to all this is that a variety of different stimuli, generated either outside or inside the body, can serve as a stressor (as defined by the release of cortisol). These include things like the physical trauma of getting punched in the nose, positive emotional stimuli such as receiving a warm smile from someone you are attracted to, or psychological stressors that are associated with moments of high anxiety (such as reading those last few paragraphs). Everyone has his or her own favorite examples of the latter.

My most familiar moments of high anxiety have changed over the years. When I was a kid, they typically involved a blissful daydream punctuated by the sound of my name being called by Nurse Ratched, our fifth grade algebra teacher who made it clear that if I couldn't expand x into a third degree polynomial, I might as well "think plastics". These days they're more likely to occur on a sleepless night at 3am, when I'm positive that the numbness in my left arm is being caused by a brain tumor.

The Autonomic Nervous System

The primary means by which your brain tells the rest of your body what to do is by sending signals through nerves that run from your brain, down the spinal cord to peripheral tissues (various muscles, glands, and other organs). These peripheral tissues reciprocate by sending feedback information back through a different portion of your spinal cord, up to your brain in order to let it know their state of affairs.

The peripheral nervous system has two distinct components (see Figure 3-3), the somatic nervous system (SNS) and the autonomic nervous system (ANS). The single function of the SNS is the control of voluntary motor movements by activating skeletal muscles. The activation of these pathways is very fast and they are also turned off quickly.

The ANS is more complicated. The ANS innervates every other tissue and organ that is controlled by the brain, including smooth muscle, cardiac muscle, and a wide range of gland cells. It is an incredibly complex task to orchestrate yet the ANS performs its duties automatically, without our voluntary control.

The ANS is subdivided into two regions as illustrated in Figure 3-3. The *sympathetic* and *parasympathetic* divisions operate in parallel, but perform very different duties. In general, the sympathetic nervous system is like the troubled teen of a family, and revs things up in your body, while the parasympathetic system acts like the responsible older sibling that tries to restore peace and keep things quieted down. They work antagonistically to one another.

For example, sympathetic activation speeds up your heart rate, while parasympathetic activation slows it down. Both systems are active to different degrees at the same time, and the subsequent output of their target organs and muscles is the sum of the activity from both divisions. This means that you can increase your heart rate by either *enhancing* the activity of the sympathetic division or *decreasing* the activation of the parasympathetic division. Likewise, you can decrease your heart rate by either reducing sympathetic activity or increasing parasympathetic activation. Make sense? So who tells the ANS what to do?

In addition to controlling the hormonal stew that floats through your veins, the periventricular hypothalamus also controls your autonomic nervous system (I told you this area was important). In the face of a stressor, cells in this region activate a highly coordinated set of both neuronal and hormonal outputs. Essential to ANS control are cells in the periventricular hypothalamus that connect to brain stem and spinal cord cells that activate the sympathetic and parasympathetic divisions.

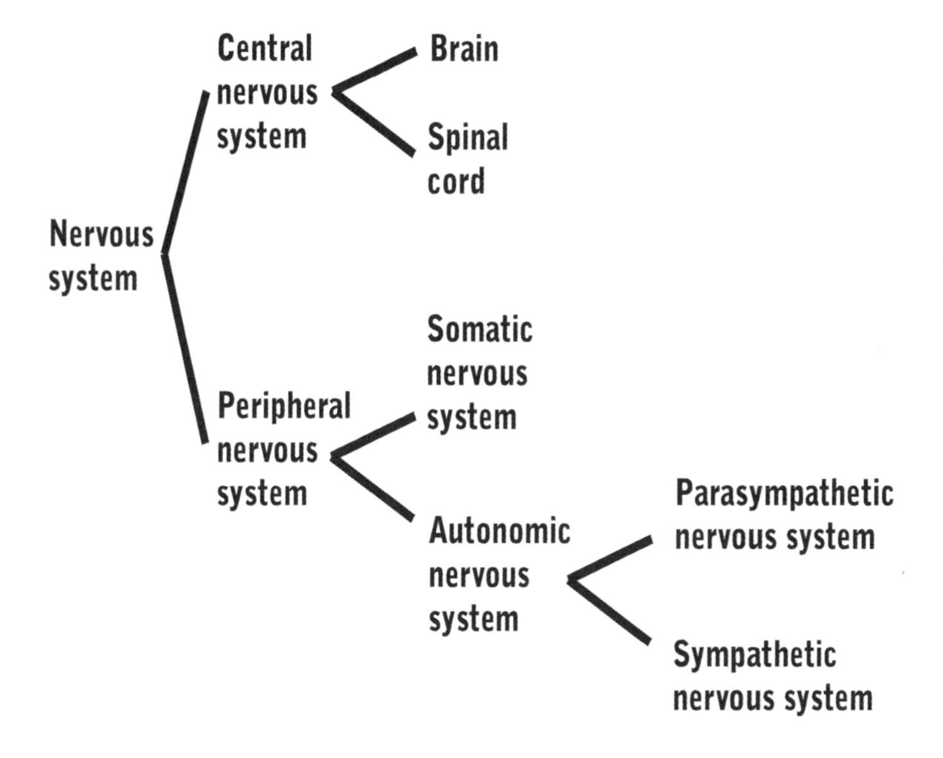

Figure 3-3 – The major divisions of the nervous system.

As shown in Figure 3-4, for the most part, cells from both divisions connect with the same set of organs. The sympathetic system tends to be most active during a crisis or stressor of some kind. Get a letter from the Internal Revenue Service requesting an audit, and you can be sure your sympathetic nervous system is screaming like there's no tomorrow. The sympathetic division is responsible for the four F's: fight, flight, fright, and reproduction. It mobilizes activity for short-term emergencies while the parasympathetic side likes to keep things calm and activate processes that are for the long-term good of the organism, such as growth, energy storage, and immune responses. The sympathetic response is for "fight or flight", and the parasympathetic response is used to "rest and digest". In most cases, when activity is high in one division of the ANS, it is low in the other, with the demands of the situation dictating which side dominates.

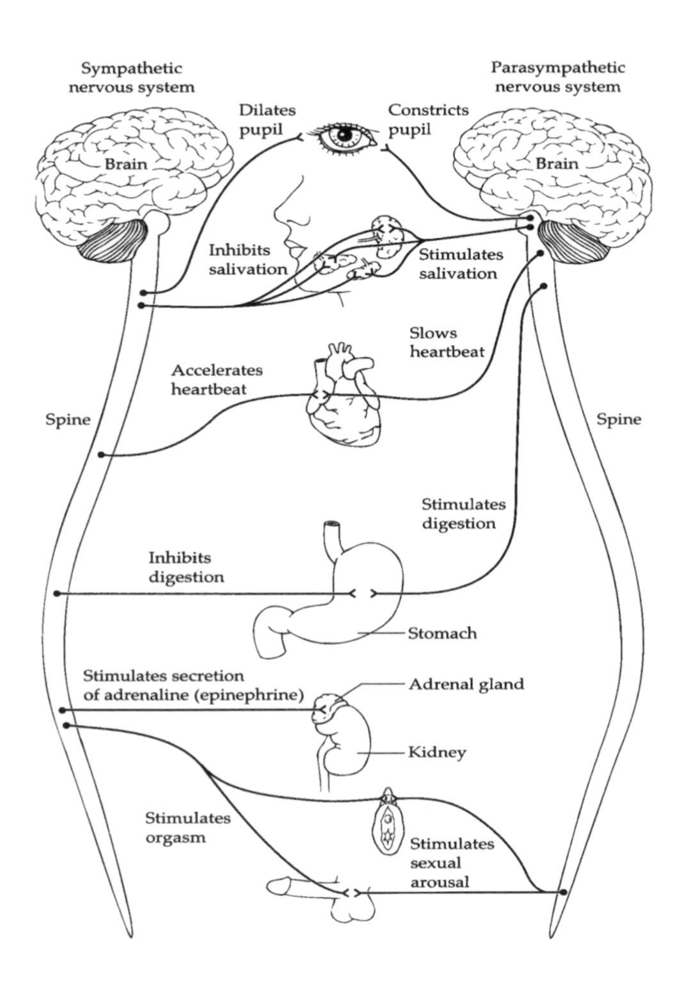

Figure 3-4 – Some of the effects of sympathetic and parasympathetic nervous system activation.

The neurons in the ANS that trigger glands to secrete, bladders to contract, airways to constrict or relax, and so on use two different neurotransmitters. Sympathetic neurons release a substance called *norepinephrine* (known as noradrenaline in the United Kingdom), which binds to receptors located on the many glands, organs, and muscles of the body where it has a predominantly excitatory effect. Another neurotransmitter, acetylcholine (ACh), is released by cells of the parasympathetic division. We will learn a great deal more about both of these transmitters in later chapters. Acetylcholine has a largely inhibitory effect on the target organs of the ANS[5].

Now remember when we were talking about the adrenal gland? We said that one effect of activating the hypothalamic-pituitary-adrenal axis is the release of cortisol from the adrenal cortex. Well, sympathetic activation causes the release of another substance, epinephrine (also known as adrenaline in the United Kingdom), but this time, from the adrenal medulla (that other part of the adrenal gland). Once released, epinephrine enters the general bloodstream, where it acts like a *sympathomimetic* (sympathetic division activator). This is where the old saying, "get your adrenaline pumping" comes from.

In summary, when a stressor is encountered, the periventricular region of the hypothalamus is the coordinating point where the brain tells the endocrine system what to do next. It orchestrates an elaborate set of neuronal and hormonal responses that quickly shift your body into a state of preparedness to deal with the crisis. One note of importance that will be re-visited later is that the cortisol and NE components of the stress response operate on two different timescales. The NE response occurs within seconds of encountering a stressor, while the cortisol response occurs only after minutes have passed and may continue for hours, long after the source of stress is gone. What is the reason for this? There are several possibilities, all of which are fairly controversial.

One idea is that the cortisol response may actually be involved in turning the NE stress response off. In this scenario, cortisol may serve a recovery role in the stress response. Another notion is that the cortisol response lasts longer because it prepares one for the next stressor. In other words, it shifts an already stressed person, into a state of increased vigilance and arousal. This creates a mechanism for *anticipatory reactions to future stressors*, and explains why many psychological stressors can produce an increase in circulating cortisol.

More importantly and relevant to our discussion, there is exciting data showing that trauma, whether psychological of physical in nature, can produce long-lasting changes in the cortisol response to stress. It's unknown how severe the traumatic event must be, and how long the change to the HPA system lasts, but some researchers estimate the alteration in the response may last a lifetime. That's right – you get super stressed out as a kid, and you may end up having an exaggerated response to all stressors for the rest of your life. In the next section, we'll explore this effect in detail. Why does this exaggerated stress response occur in some individuals and not others? How much stress does one need to experience in childhood for this change in the response to occur? There is good news and bad news, and we will discuss intriguing new findings that suggest ways in which behavioral or pharmacological treatments can combat this process. We will then explore how these mechanisms are involved in regulating mood and examine some possible therapeutic avenues for the treatment and prevention of mood and anxiety disorders that are thought to work by reducing the stress response.

Chapter 4

†

Stress and the Brain

It's probably not very difficult to find someone in your circle of friends (perhaps even yourself) who has suffered from a severe trauma early in life. This may have come in any number of forms such as suffering from child abuse (physical, sexual, or psychological), chronic child neglect (lack of shelter, nutrition, or emotional support), or loss of a parent or close caregiver. In the U.S. alone in 1995, over 3 million reports of child abuse or neglect were made, and more than 1 million were verified.

These are staggering numbers. Worst of all, individuals who experience a severe trauma early in life may be affected by this throughout adulthood, *not only in their psychological reactions to future stressors, but also in their body's physiological response*. These two components do not work in isolation, but rather interact in a complicated way. Just as we form a long-term memory of some painful event from childhood that we are able to recall in full Technicolor detail, so too does the body have a long memory. We'll see that in a wide variety of species, including humans, the experience of early life adversity can sometimes lead to alterations in the way the hypothalamus-pituitary-adrenal axis and autonomic nervous system respond to stress. Incredibly, these physiological changes can literally last a lifetime. This is important because an exaggerated stress response has been linked to an increased susceptibility for a number of disease states including hypertension, heart attack, stroke, tumor growth, the common cold, and anxiety and affective disorders to name but a few. It should give lawmakers pause to consider the relative benefits of combating early life traumas (such as abuse) in comparison to dealing with a lifetime of stress-related medical conditions.

But, there is some good news. There is intriguing research that has begun to shed light on therapeutic interventions (both behavioral and pharmacological) that can help restore the stress response to normal levels. Some of these therapies target brain sites, whereas others target peripheral portions of the HPA axis and ANS. One thing is very clear – a greater understanding of the biological mechanisms that underlie the long-term changes in stress response, will offer new potential avenues for the therapeutic regulation of mood and anxiety related disorders. The more we know about the circuitry involved in these adaptations, the better chance we have at alleviating them. After reading the last chapter, we know a little about the way the brain normally communicates with the endocrine system during the stress response. Let's now consider the way this process changes with either very prolonged or traumatic stress. We'll then relate these observations to the manner in which this system functions in people suffering from mood disorder.

Effects of Early Life Trauma on the Stress Response

There are many ways researchers can test an organism's stress response. The procedure most commonly used in humans is to expose the subject to a standardized psychosocial stressor in the laboratory. These experiments are not fun. Subjects are typically either asked to read a disturbing passage or view a film that has been prescreened to elicit a stress response in "normal" individuals. Another favorite method is the "public speaking task", in which subjects are given 10 minutes to prepare for (and anticipate) a subsequent 10-minute speech they must deliver in front of an audience. Yikes! Several independent studies have validated the efficacy of this task, showing it induces endocrine and autonomic changes reminiscent of stress[1].

Recent clinical experiments have shown that the physiological response to a psychosocial stressor is significantly increased in adult victims of childhood abuse compared to adults who have not had these adverse experiences. In an exciting landmark study, Christine Heim and colleagues working in Charles Nemeroff's laboratory at Emory University demonstrated enhanced HPA axis and autonomic responses to a psychosocial stressor in adult women with a history of childhood sexual and/or physical abuse[1]. Physical abuse, in this study, was defined as repeated abuse, once a month or more for at least a year. Women with a history of abuse had a significantly increased ACTH response to the test stressor (the public speaking task) compared to those who did not suffer childhood abuse.

The increase in ACTH is presumably a result of increased CRF secretion from the periventricular region of the hypothalamus. The logic being that an increase in CRF release into the pituitary will result in the subsequent release of ACTH into general circulation. Adult women with a history of childhood abuse also showed an elevated heart rate in response to the stressor as compared to women without abuse histories. Taken together, these findings indicate that adult survivors of childhood abuse have a persistently hypersensitive stress response. Furthermore, the persistence of this response suggests that the HPA axis has at least some form of long-term memory.

This scenario becomes even more intriguing when considering the data in relation to mood disorder. Abused women who also had a diagnosis of major depression (based on the DSM-IV) at the time of the study, showed an even larger stress response than those with abuse histories but no current depression. Incredibly, these women showed a 6-fold greater ACTH response to the stressor than control subjects matched for age and other variables such as medical history.

Abused women with current major depression also exhibited increased cortisol levels in response to the stressor, as compared to age-matched control subjects and abused women without major depression. In these women, a high degree of correlation existed between the level of ACTH and cortisol responses to stress, the magnitude of the abuse experienced in childhood, and the severity of the current major depression. Thus, women with severer depressions tended to have the largest ACTH and cortisol responses to the stressor. Interestingly, women who were abuse victims, but *did not have* a current major depression, exhibited increased ACTH responses, but had normal, or in some cases, decreased cortisol levels evoked by the stressor.

These data can be interpreted in several ways. The most parsimonious explanation (experimenters love to use this phrase) is that individuals who have a history of childhood abuse, will tend to have a hypersensitive HPA stress response, whether they currently have a major depression or not. This was demonstrated by showing both groups had an increased ACTH response to stress compared to control subjects who were not abuse victims. There are several possible reasons that ACTH may be increased in these women.

1. Cells in the periventricular region of the hypothalamus may release more CRF into the pituitary in response to stress in abuse victims, as compared to women who were not abused. Why would these cells *do* this? One scenario is that CRF-containing cells in the hypothalamus of abuse victims become sensitized, and more easily excited. This may be because more cells activate them following abuse, or perhaps it is due to alterations in an intrinsic property of the cells that make them more excitable even though the input to them is unchanged. We don't know which scenario is true. Another way of saying this is that we do not know where this form of learning is maintained neurally.

2. Another possible explanation for the increased ACTH response to stress in abuse victims is that perhaps the CRF receptors on cells in the anterior pituitary become more sensitive or "up-regulated". In this scenario, cells in the pituitary, rather than the hypothalamus, change by growing more receptors for CRF. Why would they *do* that? This could be a response to a change in the level of CRF release. If a reduced amount of transmitter is released from a presynaptic cell to a postsynaptic cell, over time, the postsynaptic cell will adapt by expressing more receptors for the transmitter. Think of this as an attempt to maintain a certain level of communication between the cells. Less transmitter is released from the presynpatic cell, so the postsynaptic cell grows more receptors so it can "soak up" what little is released. Most often, the effects counterbalance each other – decreased transmitter release results in more receptors being expressed on the postsynaptic cell, while too much transmitter being released results in a "down-regulation" or less receptors being expressed. Again, this is all an attempt to maintain a proper balance of communication between the two cells. This is the way it *usually* happens. But there are exceptions to almost everything in biology. In some cases, increased transmitter release leads to up-regulation of receptors and decreased transmitter release leads to down-

regulation. We don't yet know which occurs in individuals with a hypersensitive stress response. It may be that these folks actually have an increase in CRF release which actually up-regulates CRF receptors in the anterior pituitary. Another reason for an increase of CRF receptors in the anterior pituitary may be due to some alteration in gene expression of these proteins (the CRF receptors) that has nothing to do with changing levels of CRF secretion per se. We simply don't know if this is the case.

Are all these distinctions really important or is this just a lot of scientific hair-splitting? The critical point is that the first possibility suggests the locus of the problem is in the hypothalamus, while the second suggests it is the pituitary that changes. This distinction is very important in that different mechanisms lead to different therapeutic (e.g. pharmacological) approaches. For example, if the first possible interpretation is true – more CRF secretion in abuse victims – one way to treat this is to develop drugs that block CRF from activating cells in the pituitary. These are called *CRF antagonists* because they bind to CRF receptors in the pituitary, but they do not activate the postsynaptic cell, and therefore no ACTH is released by their actions. In other words, the CRF antagonists compete with endogenous, circulating CRF for receptors in the pituitary. Some of the real CRF will get to bind with receptors on cells in the pituitary and cause the subsequent release of ACTH from these cells. But when a drug that acts like a CRF antagonist is also present in the pituitary, many of the receptors for CRF will be pre-bound by the fake substance, and thus will not contribute to the eventual release of ACTH. The net effect is that competitive CRF antagonists reduce the amount of ACTH released from the pituitary, perhaps bringing these levels back into the acceptable physiological range observed in individuals with a normal stress response.

If, on the other hand, the second possibility is true – that abuse victims have an up-regulation (or over-expression) of CRF receptors on cells in the pituitary in response to the trauma – a different treatment course might be better. For example, in this case, either a weak ACTH antagonist or a drug that regulates the gene transcription of CRF receptors (see Chapter 1), thus preventing their over-expression, would more directly blunt the exaggerated ACTH response.

Both of these possibilities are currently being explored in research laboratories around the world. It appears from post-mortem studies and experiments using animal models (we'll get to these) that both the first and second possibilities contribute to the enhanced ACTH observed in survivors of abuse. There is evidence for *both* an increase in CRF secretion and CRF receptor up-regulation in the brains of abuse victims. This is both good and bad. It's bad news, in that it suggests more pathology to the brain circuit involved in the stress response. However, it is also good news because it allows more routes of therapy to counteract the pathology. That is, there are more points in the circuit that can be targeted by pharmacological interventions. Each of these possible targets for therapeutically restoring a normal stress response is currently being tested experimentally in both humans and animals models. We'll take a closer look shortly at the animal models, and what they do and do not tell us about the relationship between an enhanced stress response and the subsequent development of mood and anxiety disorders. For now, however, let's consider the implications of increased CRF, ACTH, and cortisol levels evoked by stress in abuse victims who also have current major depression.

Hormonal Abnormalities and Mood

There are a number of well-known changes in hormonal regulation that have been observed in people suffering from mood disorders. Depressed patients have consistently been shown to have a weakened response to several substances that stimulate the release of growth hormone. They also display a shunted response to the hypothalamic substance that normally evokes secretion of the thyroid-stimulating hormone from the pituitary. Indeed, a fairly common reason that patients are unresponsive to antidepressant treatment is the presence of a previously undiagnosed thyroid deficiency. In such individuals, administration of thyrotropin-releasing hormone has been shown to result in an acute antidepressant response. Clearly, any patient suffering from depression who is about to begin pharmacological treatment with antidepressant medications should first have a thyroid deficiency screening.

These findings are certainly interesting, but at present the strongest case for a link between hormonal abnormalities and mood disturbances is the hyper-responsiveness of the hypothalamic-pituitary-adrenal system that we have been discussing. In the 1980s and 1990s, several groups began to report increased activity in the HPA axis of unmedicated depressed patients, as shown by increased levels of cortisol in blood, urine, and cerebral spinal fluid[2]. After being given a standard dose of dexamethasone, a synthetic glucocorticoid that normally suppresses pituitary-adrenal activity through negative-feedback inhibition in healthy subjects (see Chapter 3), the majority of depressed patients failed to show suppression of ACTH and cortisol release[3]. In other words, you give dexamethasone to a healthy individual, and this person's body thinks it's actually cortisol. Remember that cortisol circulates throughout the bloodstream and eventually binds to glucocorticoid receptors in the pituitary, hypothalamus, and hippocampus that result in signals being sent to stop additional secretion of cortisol from the adrenal gland. Thus through negative feedback, the proper amount of circulating cortisol is maintained. When given the same Dexamethasone Supression Test (DST), most depressed patients fail to show a subsequent reduction in cortisol release. It's as if their bodies do not recognize that there is too much circulating cortisol, and the adrenal gland goes right ahead and keeps pumping more of the hormone out into general circulation. Consequently the negative-feedback inhibition of circulating cortisol does not function properly in depressed patients, and as you have probably deduced by now, may be one mechanism by which cortisol is increased in these folks.

Since negative-feedback of cortisol is activated by the hormone binding to glucocorticoid receptors, one theory suggests that perhaps depressed patients have either less or defective glucocorticoid receptors in these regions mediating the feedback inhibition. One indirect way to test this is by counting the number of glucocorticoid receptors expressed on cells called lymphocytes. Lymphocytes are a class of cells involved in the immune response, and are easily extracted from blood samples. In a recent study, it was shown that depressed patients exhibited a 70 percent reduction in the number of lymphocyte glucocorticoid receptor sites per cell compared with matched controls. Furthermore, depressives who were responsive to treatment with antidepressants showed an increase in the number of glucocorticoid binding sites toward the normal range observed in healthy individuals within 6 weeks of treatment.

Depressed patients have also been shown to have an increased CRF activation as evidenced by elevated levels of CRF circulating in their cerebral spinal fluid, when compared to healthy controls and patients with other psychiatric disorders. Increased levels of CRF concentrations were also observed in the cisternal cerebral spinal fluid of suicide victims, most of who were presumably suffering from major depression.

Since cerebral spinal fluid measures of CRF reflect, not only concentrations released from the hypothalamus, but other brain areas as well, it is very difficult to localize where the increased activation comes from. However, post-mortem studies can provide this information by examining the brains of deceased patients suffering from depression and measuring regional distribution (in different brain areas) of messenger RNA transcripts (mRNA) that normally control the synthesis of CRF. Such studies have already been performed and show a marked increase in CRF mRNA expression in the hypothalamus of depressed patients versus control subjects. These studies have also shown an increase in the number of CRF-containing neurons in the hypothalamus of post-mortem depressed patients versus matched controls. This indicates that depressed patients show an increase in hypothalamic CRF secretion, which in turn, may contribute to increases in ACTH and cortisol levels.

Hundreds of studies since these early pioneering experiments have replicated the initial demonstrations of HPA axis hyperactivity in patients suffering from major depressive and anxiety disorders. Indeed, this finding is perhaps the most highly replicated in all of biological psychiatry. Additional investigations have now revealed changes at various points along the HPA axis in patients suffering from mood and anxiety disorders. For instance, in addition to increased circulating levels of CRF and cortisol, many patients with mood disorders actually have enlarged pituitary and adrenal glands relative to healthy control subjects. During depression, approximately 70 percent of patients exhibit adrenal gland enlargement of 1.7-fold versus matched controls. This means the adrenal glands of these patients almost double in size. Interestingly, the enlargement subsides upon successful treatment with antidepressants. Enlargement of the adrenal gland may be one contributing factor to increased cortisol release (from the adrenal gland) observed in depressed patients.

In addition to abnormal swelling of the adrenal gland, many depressed patients have enlarged pituitaries, which may be mediated by an increase in CRF release into the pituitary from the hypothalamus. We've already mentioned that depressed patients show elevated CRF release. The increase in CRF release and the enlargement of the pituitary appear to be *state-dependent rather than a trait marker*, since they return to normal levels with successful treatment of depression. Several studies have now confirmed that pituitary and adrenal gland enlargement, as well as increased CRF and cortisol secretion are all present during a depressive episode, but return to normal levels following a similar time course to the alleviation of the depressive symptoms as indicated in self reports and the Structured Clinical Interview for DSM-IV.

Stress and mood: cause or effect?

Well, this is all very interesting, but several important questions remain. For instance, are these changes in HPA axis function *the cause or the effect* of mood disorder? There are a number of findings that suggest the changes may be a causal factor.

First, people with medical conditions such as Cushing's syndrome, in which different types of tumors cause a vast over-secretion of cortisol, become clinically depressed and terribly anxious. The most commonly described psychiatric manifestations of the disease include anxiety (79%), depression (68%) often with comorbid anxiety, and panic disorder (53%).

Laurie, a Cushing's syndrome patient recently observed, "it's like a parade of ants marching up and down your spine, making it tingle, but there is no way to get rid of them". Cushing's syndrome is treated by a number of procedures including administration of metyrapone, a glucocorticoid synthesis inhibitor, adrenalectomy, or pituitary irradiation or resection, each of which reduces circulating cortisol levels and subsequently the mood and anxiety disorder symptoms. Metyrapone is currently being tested as an antidepressant agent in patients without Cushing's syndrome and in preclinical trials using animals. So far, the results are encouraging, yet still preliminary.

A second observation suggesting HPA axis hyperactivity may be a cause rather than an effect of major mood and anxiety disorders comes from people with any number of diseases who are prescribed glucocorticoids (steroids) for treatment purposes. Steroid treatment is common in a variety of disorders ranging from asthma to multiple sclerosis. Many patients treated with steroids become clinically depressed, and in these people, there is a significant positive relationship between the dosage level and the severity of the depression.

Of course, this is still a bit tricky to understand. Do these people become depressed because of an increase in glucocorticoid circulation or because they recognize they have a serious medical condition? You can show that it is the increase in glucocorticoid levels that is linked to the depression by comparing people with the same disease and severity who are not receiving glucocorticoids as treatment, but rather some alternative. In such studies, depression rates are highest among those patients being treated with glucocorticoids when compared to controls matched for disease severity and duration. This implies glucocorticoids may play a causal role in the development of mood disorders. Finally, there are many studies using animals that suggest a causal relationship between HPA axis hyperactivity and the subsequent development of depression. We turn to those now.

Animal Models of Mood Disorder

For many people, it's probably difficult to imagine how mood, a phenomenon involving higher cognitive functions such as emotional processing and memory, can be studied using non-human primates and in many cases rodents. This is part of a larger issue concerning what it is we can learn about the human condition by experimenting with animals. At every conference I attend, and every major university I have had the pleasure of working, this question is a hotbed of controversy among lay people, academics, and scientists alike. Indeed, some of the staunchest critics of animal research I know happen to be scientists who often utilize animals in their own research. This goes to show that it is not an easy question to answer, even for those engaged in the work. There is also a whole host of ethical issues that must be contended with when addressing this question. These are well beyond the scope of this book, but are critically important to consider when examining the role science (especially medical science) plays in society.

No matter which side you come down on this debate, and often it involves an "it depends…" in there somewhere, it's important to realize that once an *accurate* animal model of a disease state is developed, a process of rational investigation of therapies can begin. I use the term "rational" to distinguish this approach from the often serendipitous findings of effective drug therapies that have to date been the norm in treating psychiatric disorders. Some of these were mentioned in the introduction.

Of course, the problem is in the semantics – it is not always easy for scientists studying disease processes to agree on what is indeed an "accurate" model. Numerous animal models of depressive syndrome (or some aspects of the syndrome) and anxiety disorders have been developed and evaluated using a variety of criteria, including origin of symptoms, response to treatment, and underlying biochemical changes. Both genetic and epigenetic models have been developed.

The genetic models usually involve crossbreeding different strains of mice (transgenic approach) or selectively deleting genes that code specific proteins ("knock-out approach) that are thought to contribute to a specific neurotransmitter system or behavioral phenotype. For instance, one transgenic mouse strain that has been developed results in the overexpression of CRF. These mice produce too much CRF as compared to wild-type mice that do not have this genetic manipulation. Sound familiar? This is supposed to emulate the exaggerated CRF response to stressors seen in many patients suffering from clinical depression and stress disorders.

Another genetic model consists of the selective "knock-out" or deletion of the genes that normally code the production of glucocorticoid receptors. The logic here is that these mice do not have the glucocorticoid receptors that can be used to facilitate the feedback inhibition that normally shunts corticosterone (the cortisol equivalent in rodents) production from becoming too great. This, you'll recall is thought to be a mechanism that malfunctions in many depressed patients (as shown by a lack of response to the Dexamethasone Suppresion Test), and leads to a marked increase in cortisol levels.

So are these mice depressed? Do they exhibit any of the following symptoms described in the DSM-IV:

1. a persistent depressed mood
2. diminished interest or pleasure in all or most daily activities
3. significant weight loss or gain or marked change in appetite
4. significant change in sleep cycle, duration or quality
5. physical slowing or agitation

6. fatigue or loss of energy
7. feelings of worthlessness or inappropriate guilt
8. difficulty thinking or concentrating
9. recurrent thought of death or suicide

Well, certainly some of these markers are more easily demonstrated in animals than others. Numbers three through six can be tested in a fairly direct fashion using a variety of well-established behavioral assays. Believe it or not, the first two markers, "a persistent depressed mood" and anhedonia (lack of an ability to experience pleasure) have also been tested using a number of extremely creative behavioral measures. For instance, experimenters often take advantage of the fact that rodents enjoy the taste of sugar (this is probably a universal trait across species, don't you think). They can then grant an animal a choice between, say, two small volumes of water, one that has been sweetened with glucose and one that has not. As you can imagine, when given this choice, most rodents prefer the sweetened solution and spend more time drinking it rather than the plain old water. This is usually thought to reflect a sense of pleasure the rat is engendered with by the taste of the sweetened solution rather than a strict biological imperative for glucose consumption. Of course, it *could* be both. From an evolutionary perspective, one can imagine that the latter might lead to the development of the former. The biological requirement for glucose (to drive the many energy demands of the body) may ultimately be expressed as a preference for glucose containing foods.

Number eight can also be tested using standard learning tasks employed in most animal behavior laboratories. However, numbers seven and nine are awfully tricky. How can we possibly demonstrate that a non-human animal is thinking about suicide? I don't know, but if you have any ideas I'd love to hear them.

Suffice it to say, animal models of depression, have been shown to capture only a portion of the symptoms of this disorder. They are clearly lacking in the subtle nuances of what most of us think of as the leading symptoms of the illness, but nevertheless some of them do capture many of the manifestations that are testable in animals. Moreover, many of the models exhibit a suppression of these symptoms after being treated with classical antidepressants like flouxetine. Indeed, antidepressant treatment at least partially ameliorates the major behavioral, neuroendocrine, and neurochemical aberrations thought to contribute to mood disorder in the two genetic models mentioned above. Prozac for mice, now we're on to something!

Early adverse experiences and the adult stress response

There are several paradigms used by researchers to simulate, in animals, the effects of early life trauma to humans. These typically involve exposing a developing animal (often a rat or monkey) to a mild stressor such as a brief period of isolation rearing, maternal separation, variable foraging demands that make their food supply less predictable, or other conditions that induce a sense of "learned helplessness" or a loss of control. Such manipulations to immature animals, even for periods as brief as two weeks, have a profound impact on their stress response later as adults.

Harry Harlow of the University of Wisconsin was among the first to demonstrate the importance of the relationship between the primary caregiver and its offspring during the early stages of life. In keeping with psychological theories expounded by Freud and later James Bolby, Harlow maintained that the development of a consistent attachment to the primary caregiver plays a central role in the future development of personality and psychopathology in the offspring.

During two decades of research beginning in the 1950s, Harlow and his colleagues provided dramatic evidence for this by studying the effects of maternal separation on infant rhesus monkeys. Monkeys raised in partial isolation (i.e. raised with peers but no mother) for the first six months of life exhibited several pathological behaviors as adults that included heightened fear and aggression, an inability to cope with daily stressors, impaired social behavior with peers, learning disabilities, and a number of psychomotor alterations.

Neuroanatomical studies of monkeys who underwent the same manipulations revealed marked changes in several brain regions as compared to monkeys who did not experience partial isolation. Consistent with our theme, follow-up studies using the same paradigm adopted by Harlow showed that adult monkeys who were maternally deprived as neonates displayed increased anxiety and elevated CRF concentrations when compared to monkeys who were not deprived. To me, these findings are at the same time both unsurprising, yet astonishing. Unsurprising in the sense that it should be obvious that even a brief loss of interaction with a primary caregiver can cause great stress to an offspring. Yet it is astounding how profound an impact this form of psychological stress has on the offspring as it grows into adulthood. This brief trauma results in significant behavioral anomalies, gross alterations in brain anatomy, and hyperactivation of the HPA axis to daily stressors *years later in the adult.*

Attempts to reverse the effects of maternal separation through structured socialization with the mother and peers have met with limited success. The separated monkeys were able to cope reasonably well under low-stress conditions, and showed normal basal levels of CRF. However, they became significantly more agitated than normal monkeys when faced with a psychosocial stressor, and consistent with this response, exhibited elevated levels of stress-evoked CRF. Similar observations have been made in rodents. So in a very real sense, the animal models are mimicking the observations demonstrated in human survivors of childhood abuse.

Paul Plotsky, director of the Stress Neurobiology Laboratory at Emory University, working in conjunction with Michael Meaney, director of the Developmental Neuroendocrinology Laboratory at McGill University, performed a series of pioneering studies examining the brain mechanisms that become altered in neonates to produce an exaggerated stress response in adults. Adult rats that were separated from their mothers for 180 minutes per day on postnatal days 2-14 exhibited an increase in CRF, ACTH, and corticosterone responses to a variety of stressors in adulthood. Here, much like the observations made in monkeys and humans, we again see a hyperactive HPA axis response to a stressor in adults who underwent trauma during the early stages of development.

It appears from the work of Plotsky, Meaney, and their colleagues, that the exaggerated HPA axis response may, in part, by driven by persistent changes in CRF-containing cells in the hypothalamus. This "persistent change" in the way the hypothalamic CRF-containing cells respond to stress, is an example of brain plasticity. Thus it can be considered a type of long-term memory, albeit a form that we do not have conscious awareness of. Psychologists refer to this kind of memory as "implicit", since we do not have direct or explicit conscious control over it.

Experiencing significant stress as a child shifts these cells into a heightened mode of activation, in which even a weak stressor produces a large CRF response. The elevated CRF release then launches a cascade of events – increased ACTH release from the pituitary, leading to increased corticosterone release from the adrenal gland, and ultimately the behavioral manifestations of stress.

Among the behavioral abnormalities seen in animals with elevated HPA axis responses to stress are several quintessential features of mood disorder in humans, namely anxiety, decreased libido, insomnia, changes in sleep-cycle architecture, and decreased appetite to name just a few. Delivery of CRF directly into the brains of rats (which for ethical reasons is impossible to do with humans) produces very similar behavioral responses. These data may help explain why one of the most consistent findings in the mood disorder literature is that parental loss early in life dramatically increases the risk for depression years later.

A quite remarkable finding from this work is that the heightened stress responses at each level of the HPA axis are reversed by treatment of the maternally separated rats with antidepressants such as the selective serotonin reuptake inhibitor, paroxetine, or the selective norepinephrine reuptake inhibitor, reboxetine.

A big question, then, is if these agents are *selective* for either serotonin or norepinephrine transmission systems, how can they affect CRF, ACTH, and corticosterone release? The answer is that both the HPA axis and autonomic nervous system are highly interconnected with other brain regions that primarily utilize these transmitters. A related and interesting aspect of this research is that the increased sensitivity of the adult HPA axis to stress, appears to be limited to psychological stressors (i.e. novel environment, air puff startle, brief restrain), since it is not observed after physical stress such as being exposed to cold temperatures or a weak tail pinch. This finding suggests that the corticolimbic system, an interconnected set of brain regions implicated in memory formation and the processing of emotions (as well as many other functions), is involved in mediating the heightened stress response. It is in these "extrahypothalamic" systems that we now turn to understand how drugs or behavioral therapies that alter glucocorticoid release may subsequently affect monoamine (e.g. serotonin, norepinephrine, epinephrine, and dopamine) transmission and perhaps mood and anxiety disorder symptoms.

What do Monoamines and Stress Hormones have in Common?

Okay, by now after reading so much about stress, anxiety, and depression, you're probably wishing you purchased that John Irving novel you were looking at instead of this book, right? Just hang in there, some good news is right around the corner. In the last few sections, we discussed several observations that suggest a strong association between stress and mood disorders. We then examined how the stress response is controlled by the HPA axis and the autonomic nervous system, and some of the major ingredients of the hormonal soup that is involved in this sequence of events. From these data, it appears there is a growing body of evidence implicating stress and excessive CRF and glucocorticoid release as a *cause* of anxiety and mood disorders, rather than merely a consequence. If this is really true, then it is possible to imagine a whole new class of therapeutic interventions (both pharmacological and behavioral) that may have antidepressant and anxiolytic actions.

For those individuals who are suffering from major depression and show an elevated HPA axis response, one pharmacological treatment might consist of drugs that lower glucocorticoid secretion from the adrenal gland. Such drugs, called "adrenal steroidogenesis inhibitors", already exist and have indeed in the last few years been shown to alleviate symptoms of mood disorder in some individuals.

However, this approach is fairly tricky. As you have probably gathered by now, if you lower glucocorticoids too much, a variety of side effects may occur since several metabolic and cardio-pulminary processes may be altered. Yet it is incredibly exciting that anti-glucocorticoids can, in some instances, serve as antidepressants. This early research may engender the development of a new class of pharmacological agents that better position the exaggerated glucocorticoid response back into a normal physiological range that reduces mood disturbances while avoiding side effects.

Similar trials are being conducted with drugs that selectively block CRF from binding to its target receptors. One drug in clinical trials called R121919 (I'm sure the manufacturer will think of a sexier name before it hits the market), binds to CRF receptors in the pituitary, preventing the actual CRF from binding at these sites. (Remember, this is called an "antagonist".)

The first test of this drug on mood disorder in humans was recently completed at the Max Plank Institute of Psychiatry in Munich, Germany with impressive results[4]. The test was originally performed to determine if the endocrine actions of the drug compromise the stress response system or if other safety and tolerability issues exist. The compound was administered to 24 patients suffering from major depressive disorder, and was found to be well tolerated by most patients in the observation period and did not impair baseline ACTH and cortisol secretion under low or no stress conditions.

This means that side effects associated with an attenuated HPA axis that are only active during non-stressful situations should be minimized. This is great news, but even more interesting is the fact that significant reduction in depression and anxiety scores (using both patient and clinician ratings) were observed. These reductions were maintained until patients discontinued drug administration, at which point some experienced a worsening of depressive symptoms.

Taken together, these data suggest that CRF-receptor antagonism may have considerable therapeutic value in treating major depression and perhaps anxiety disorders that involve an exaggerated CRF response at baseline conditions (low or no stress) or following stress exposure.

So here we have data from laboratory animals and humans suggesting a link between an exaggerated stress response and the subsequent development of mood disorder symptoms. There is also recent data indicating that glucocorticoid or CRF antagonism has antidepressant actions in humans. "But wait a minute", you're thinking. "Didn't I hear somewhere that there is supposed to be a direct connection between levels of monoamine transmission, particularly serotonin and norepinephrine, and mood disorder? What do these transmitter systems have to do with CRF and glucocorticoids?"

It's true that for perhaps the last forty years or so, the monoamine deficiency explanation has been the leading biological theory of depressive illness. Many of the chief discoveries up until the last decade point to a serotonin and/or norepinephrine reduction as being *the* causal factor in depression. After all, that's how antidepressants work isn't it – by blocking the reuptake of these two transmitters and thereby increasing their concentrations in the synapse? If only it were that simple.

We think SSRIs and SNRIs do increase the levels of their respective transmitters in the synapse. We *think* they do. However, it's not clear if sustained use of these reuptake inhibitors results in a consistent and sustained increase in transmitter concentration in the synapse or whether this increase is just transitory.

Biological systems love to compensate for changes in normal functioning. If transmitter levels are increased in the synapse for a prolonged period of time, you can bet that other adaptive changes will begin to occur, most notably at the receptor level. For instance a down-regulation of receptors on the postsynaptic cell may occur to compensate for the increase in transmitter. That is, the postsynaptic cell sees more transmitters floating around than normal because reuptake back into the presynaptic cell is being blocked (by an SSRI or SNRI). Consequently, over time it starts to shut down some of the receptors for the transmitter in an effort to maintain a specific range of communication between the cells. It's kind of like the postsynaptic cell, says, "stop shouting, I can hear you!"

The next logical question to ask, then, is where exactly is the mechanism by which SSRIs and SNRIs alleviate depressive symptoms? This is perhaps the most difficult question to answer. Many researchers in the field, including the author, think it is unlikely that the wide range of distinctly different symptoms comprising a typical mood disorder can be accounted for by disturbances in a single transmitter system (e.g. serotonin), or perhaps even in a single *class* of transmitter systems (e.g. the monoamines). And, as mentioned above, none of these systems are really independent or operating in isolation. As we'll see, there are extensive connections between brain regions that mediate the stress response and those that utilize monoamines as their primary transmitter. Indeed, based on the vast array of behavioral symptoms of mood disorder, one must expect that a number of brain areas are involved in the pathology.

The complicated way these systems interact and our inability to currently understand exactly how they change during affective disorders, has created a great deal of confusion and controversy both within scientific circles and for the general public. For instance, we know that glucocorticoids can alter many features of monoamine neurotransmission including the amount of transmitter synthesized, how quickly the transmitter is metabolized or taken back up into the presynaptic cell, how many receptors for the transmitter are expressed on both the pre- and postsynaptic cells, the efficacy of the receptors, the subsequent changes that occur in the postsynaptic cell after a receptor is activated, and so on.

As will be discussed shortly, the monoamines, in turn, can influence stress hormone circulation. In scientific circles, attempts at relating either monoamines or stress hormones to the etiology of mood disorder can sometimes lead to a "which came first, the chicken or the egg' brand of theorizing. Do increased levels of circulating stress hormones lead to alterations in monoamine transmission or is it the other way around? Finally, are both critical components that cause mood disorder symptoms, or is one epiphenomenal?

The classic interpretation is that major depression is stressful, thus raising glucocorticoid levels. Once the depression is treated with antidepressants, monoamine levels are reduced, alleviating the depression, which makes life less stressful, and results in a decrease in stress hormone levels.

A more radical interpretation, but one that is gathering steam within the scientific community is that the etiology of some mood disorders begins with an exaggerated stress response. This may be a consequence of early adverse life conditions like those we have been discussing, a genetic predisposition toward a heightened HPA axis response, or some combination of the two. The theory suggests that the increased levels of CRF and glucocorticoid secretion *lead* to subsequent changes in monoamine transmission and mood disturbances.

In this scenario, antidepressants work by reducing glucocorticoid and/or CRF levels back into a normal physiological range, causing a normalization of monoamine transmission and alleviating the depression. As mentioned earlier, there is some evidence showing that antidepressant treatment reduces glucocorticoid and CRF circulation. Clearly, more research is needed to understand how the monoamine and stress response systems interact in order to locate the mechanism of the effect. It should also be pointed out that if indeed an exaggerated HPA axis response is a cause of mood disturbances, it is certainly not the cause in *all* depressions. Estimates are that only 55-65% of people suffering from major depression have an exaggerated HPA axis response, either at baseline or after being exposed to a stressor. This indicates that there are probably several causal routes to a mood disorder and the HPA axis impairments may be only one of them.

Just as likely is the possibility that the HPA axis and monoamine systems become less coupled in some folks who have a mood disorder, and dysfunction becomes more restricted to one system or the other. This may be related to why some patients, but not others, exhibit symptoms of anxiety disorder in conjunction with major depression. To get a better feel for how this may occur, let's look at the points where these two systems talk with each other.

Chapter 5

†

The Broader Picture: Stress Influences the Monoamine Systems

One of the main points of the last two chapters has been that psychological stressors, which can come in a variety of guises - a feeling of helplessness or loss of control, separation from a primary caregiver, unpredictability in the environment – have a profound impact on our stress response that persists into adulthood. Moreover, the activation of the stress response system in both humans and laboratory animals elicits a battery of symptoms at the physiological, behavioral, and cognitive levels that resemble several elements of mood disorder. It is very exciting, then, that research has now begun to elucidate how the stress response system can have a major influence over widespread regions of the brain, including systems that use monoamines as there primary neurotransmitter. Such information has granted us an opportunity to develop a deeper and more holistic understanding of the biological antecedents of mood and its disorders. For this to happen, however, we must get beyond the dogma of the prevailing theories.

To begin, we need to determine which brain systems are involved in the processing of emotions and mood disorders, and exactly how they influence each other. Let's discuss what research has shown thus far.

How the Stress Response System Talks with the Rest of the Brain

In addition to the periventricular region of the hypothalamus, cells that contain CRF are also located throughout the corticolimbic system. These sites include the cerebral cortex, the amygdala, and the brain stem. We have already mentioned the cerebral cortex (in Chapter 3) and its role in numerous higher cognitive functions. The amygdala has been implicated in multiple functions related to the processing of emotions, and the brain stem contains several discrete groups of cells (referred to as nuclei) that support basic regulatory functions such as levels of arousal, sleep-wake cycles, digestion, respiration and others.

An important function of the brain stem is that it serves as the origin of several cell groups that synthesize and use monoamines as their primary neurotransmitter. In particular, the origin of the serotonin system is a group of cells in the brain stem called the raphe nuclei, and the origin of the norepinephrine system is comprised of several cell groups, most notably the locus coeruleus. These nuclei project to various areas throughout the brain.

Thus CRF not only operates as a hormonal releasing factor (in the endocrine system), but also as a neurotransmitter at several locations throughout the corticolimbic system in its response to stress.

I have shown in Figure 5-1, a schematic of how the HPA axis and the autonomic nervous system interact with the corticolimbic system during the stress response. So far we have discussed the HPA axis illustrated on the left side of the figure. Stress activates the hypothalamus, which releases CRF into the pituitary, initiating the endocrine response to stress. This eventually results in increased levels of circulating cortisol concentrations.

In addition to regulating HPA axis activity, cortisol activates the amygdala, which, in turn, is involved in initiating the autonomic nervous system response to stress and releasing CRF into the locus coeruleus – the birthplace of the norepinephrine system.

The locus coeruleus continues the group conversation by communicating with a number of different brain regions including the periventricular hypothalamus (where this all started), the amygdala, the cerebral cortex (particularly the prefrontal cortex), and other areas in the brain stem (Figure 5-1).

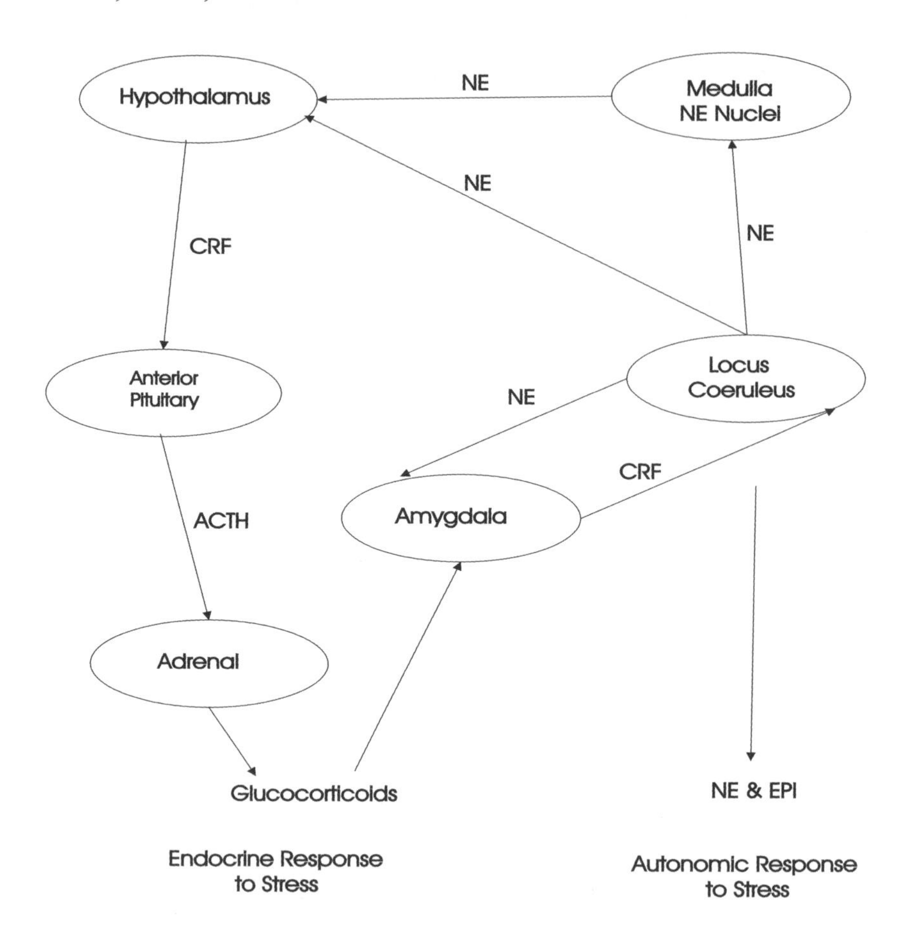

Figure 5-1 – The HPA axis interacts with the norepinephrine (NE) system to launch the autonomic response to stress and to communicate with several brain structures. Stressful stimuli activate cells in the periventricular hypothalamus, which initiates the endocrine response to stress by activating the HPA axis (shown on the left side of the figure). There are several brain areas influenced by the release of glucocorticoids that result form HPA axis activation. Glucocorticoids that bind to receptors in the pituitary and hypothalamus result in a negative feedback of future release of the hormone (not shown). Glucocorticoids also activate the amygdala and induce the release of CRF into the locus coeruleus, a brain area rich in norepinephrine-containing cells.

> When activated by CRF, the locus coeruleus releases NE back into the amygdala (completing a positive feedback loop), the hypothalamus (another positive feedback loop involving the entire HPA axis), and various nuclei in the brain stem.

Okay, so let's try to decipher all this mess. Many of you are probably thinking that somehow the term, "quagmire" applies. It's true, the general circuitry involved in the stress response is complicated, but there are several key features that can help simplify our understanding of this process and how it may malfunction and contribute to anxiety and mood disorders.

The first key is to look for all the feedback loops. For instance, once the is activated and releases CRF into the locus coeruleus, the locus coeruleus says, "thank you very much, here is some norepinephrine". The appearance of norepinephrine in the amygdala causes it to release yet more CRF into the locus coeruleus, which continues the positive feedback cycle on and on. The release of norepinephrine from the locus coeruleus into various areas of the brain mediates the autonomic nervous system response to stress.

So, once the HPA axis is activated, cortisol binds to cells in the amygdala, which, in turn, initiates the ANS component of the stress response through the release of norepinephrine from the locus coeruleus. What happens next, you ask?

Once the locus coeruleus is activated in this process, it releases norepinephrine into the prefrontal cortex and periventricular hypothalamus. And, as you may have surmised, the increase in norepinephrine in the hypothalamus promotes the additional release of even more CRF, to start the HPA axis activation all over again. This is yet another positive feedback loop between norepinephrine and CRF. When you increase either norepinephrine or CRF in one of these loops, you automatically get an increase in the other substance. Likewise, when you decrease one substance, you get a subsequent decrease in the other.

Can you see all these loops? *The take home message is that norepinephrine promotes the synthesis and release of CRF, and CRF, in turn, promotes the synthesis and release of norepinephrine.* However, it should be noted that this is not the sole function or effect of norepinephrine in the central nervous system.

Norepinephrine is part of a diffuse modulatory system that extends to large portions of the brain, and we are only beginning to understand some of the functional properties of these circuits and how they change as a result of stress and/or early life trauma. As might be expected from the anatomical connections, the norepinephrine system, much like the HPA axis, can become hypersensitive as a result of stress experienced either in childhood or as an adult.

Monkeys that are stressed as neonates due to brief periods of maternal separation, exhibit hyperactivation of the locus coeruleus as adults when compared to non-deprived monkeys. This increased activation implies a hypersecretion of norepinephrine from the locus coeruleus to the target sites shown in Figure 5-1.

One important effect of the increased release of norepinephrine, is that more CRF will be secreted from the hypothalamus and amygdala, thus producing yet more norpinephrine release from the locus coeruleus, and before you know it, you have a self-sustaining cycle that can become even worse with the smallest of stressors.

When adult monkeys that have experienced trauma early in life are given even a small dose of yohimbine, a substance that facilitates the release of norepinephrine from the locus coeruleus and other sites, they exhibit several characteristic symptoms of anxiety and affective disorder shown in laboratory animals and humans with excessive CRF or glucocorticoid secretion. These symptomatic behaviors are not observed in monkeys given the same dosage of yohimbine that did *not* experience early life trauma. These results have since been replicated in humans.

This is quite interesting because it suggests severe stress experienced early in life can rev up our endocrine and autonomic responses to future stressors as adults. It also implies that some components of monoamine transmission are persistently altered into adulthood. This type of subtle alteration in norepinephrine secretion and binding may conceivably result in a host of functional changes that include a propensity toward anxiety and mood disorders, sleep and appetite disturbances, and sexual dysfunction to name just a few from a very long list. And indeed, many of these problems may not be experienced until a stressor is present.

Experiments show that even very weak, but chronic stressors can induce behavioral symptoms in laboratory animals that have a hypersensitive norepinephrine system. As humans, we're usually smart enough to avoid very large stressors. We have laws and cultural mores that seek to minimize such possible threats. Consequently, most of us enjoy mild stressors, but lots of them. Traffic jams, problems with co-workers or significant others, politics, financial burdens, you name it – the list goes on and on. We'll see in Chapter 10 that many of the brain regions involved in this circuit, are altered in primary mood disorders.

Serotonin and stress

So what about serotonin? The involvement of serotonin circuits in the stress response is even more complicated (sorry) and less understood than the norepinephrine circuits. Let's begin with some basics.

There are direct anatomical connections from the serotonin system to the HPA axis. The origin of the serotonin system is a series of nine cell groups in the brain stem, collectively referred to as the raphe nuclei. When serotonin-containing cells in the raphe nuclei become active, they release serotonin into a number of target areas involved in the stress response including the amygdala, periventricular hypothalamus, prefrontal cortex, and the hippocampus. This is a familiar cast of characters, right?

Although the anatomical targets are similar to the norepinephrine system, the serotonin system has a more complex relationship with the HPA axis. For instance, there is now evidence that glucocorticoid and serotonin receptors interact functionally. For instance, if you give a laboratory animal an injection of serotonin, the typical response is an increase in the number of glucocorticoid receptors in several regions of the limbic system including the hippocampus. This is thought to result from a decrease in glucocorticoid circulation (Figure 5-2)[1].

The control of this functional relationship is two-way, similar changes in serotonin receptors have been reported with manipulations of glucocorticoid levels. Working at the Mental Health Research Institute and Department of Psychiatry at the University of Michigan, Juan Lopez, Stanley Watson, and their colleagues have shown that chronic, unpredictable stress (a putative animal model of depression) which produces high levels of circulating glucocorticoids in the rat, leads to a reduction in serotonin receptor binding in the hippocampus[2]. This is consistent with the idea that *an exaggerated HPA axis response to stress may, under chronic conditions, result in significant reductions in serotonin receptor activation*[3].

The reduction in serotonin receptor binding in this model is prevented by removing the adrenal gland, indicating that the effect is indeed, mediated by increased glucocorticoid levels (since glucocorticoid is released from the adrenal gland). Interestingly, the reduction is also prevented by the administration of tricyclic antidepressants.
There are now a number of reports suggesting that similar mechanisms may be at work in humans. For example, suicide victims with a history of depression often have decreases in serotonin receptor mRNA in the hippocampus. This finding is consistent with observations that suicide victims and patients with a history of chronic depression sometimes display a hyperactive HPA axis as indicated by enlarged adrenal glands, pituitaries, and elevated cisternal CRF. *Taken together, the laboratory animal and human data suggest a close link between an overactive HPA axis response and abnormalities in normal serotonin release and binding.*

So what are we to make of the interactions between the HPA axis and the monoamine transmitter systems? Simply put, alterations in the functioning of one system affect the functioning of the other.

Persistent changes in HPA axis responses to stressful stimuli result in long-lasting changes in monoamine release and receptor binding. Conversely, long-term alterations in monoamine system functioning contribute to stable pathological changes in the HPA axis response to stress. This opens up new possibilities for treating clinical syndromes such as mood and anxiety disorders associated with impairments to either system. If hyperactivity of the HPA axis is a contributing factor to the changes in norepinephrine and serotonin transmission seen in major depression, treatments that reduce CRF, ACTH, or glucocorticoid levels may weaken the precipitating effects of stress on this disorder. The efficacy of such treatment possibilities is currently being tested in a number of clinical trials, and as mentioned earlier has resulted in encouraging, albeit preliminary results. Stay tuned to further developments by checking the web sites listed in the appendix for up to the minute results of on-going and new clinical trials.

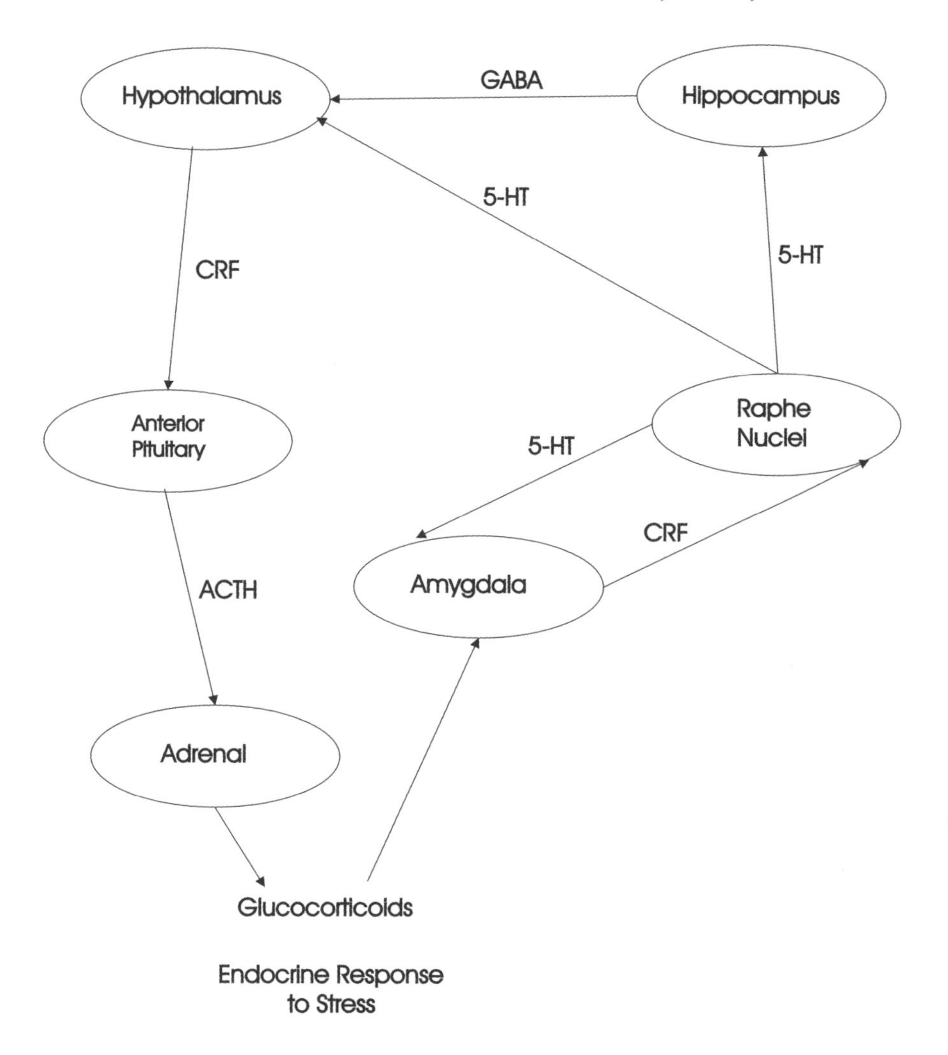

Figure 5-2 – The HPA axis interacts with the serotonin (5-HT for 5-hydroxytryptamine) system at a number of corticolimbic sites. Activation of the HPA axis by a stressor results in the secretion of glucocorticoids from the adrenal gland. Glucocorticoids cause the amygdala to become activated and release CRF into several cell groups including the raphe nuclei. The raphe nuclei are a series of nine cell groups, rich in serotonin-containing cells that have a normal baseline level of activity that is reduced when CRF is present in this region. Thus an exaggerated

stress response that results in a hyper-secretion of glucocorticoids into general circulation typically leads to a reduction in serotonin release and receptor binding. Serotonin has an inhibitory control over the HPA axis. It is believed that serotonin activates inhibitory cells in the hypothalamus, hippocampus, and several other regions indirectly through the hippocampus, the lateral septum, and the bed nucleus stria terminalis (the latter two not shown). This set of connections and their transmitter properties leads to an antagonistic relationship between HPA axis activation and serotonin release and binding. Generally speaking, increases in glucocorticoid or CRF secretion attenuate the serotonin system and vice versa. This suggests that pharmacological or behavioral manipulations that reduce HPA axis activity may have an effect of enhancing serotonin release and binding. Such manipulations may have clinical relevance for the treatment of mood and anxiety disorders.

Chapter 6

†

Birth and Death of a Brain Cell

As we have seen, severe stress experienced early in life can have a profound impact on the way our body responds to stress in adulthood. Such experiences can lead to long-term, if not permanent, hyperactivity of the HPA axis and norepinephrine system. They likely lead to persistent changes in other monoamine systems as well, namely those circuits using serotonin as their primary transmitter. This collection of modifications can be thought of as a sensitization of the circuits mediating the stress response, in that even very mild stress in adulthood can lead to an exaggerated response.

Many of these alterations are produced by rather subtle changes in the release and binding of the hormones and transmitters we have been discussing. However, there is now evidence demonstrating that excessive HPA axis and monoamine transmission can also lead to gross abnormalities in brain cell structure, decreases in neurogenesis in certain regions of the corticolimbic system, and even cell death.

Much of the work showing these effects has been done in the hippocampus, a brain region we study in my laboratory. The hippocampus, as shown on the bottom of Figure 1-2, is located in the medial temporal lobe, and has been implicated in a number of cognitive functions, including learning, memory, and the processing of emotions. This region also contributes to several pathological states including epilepsy, schizophrenia, anxiety, and mood disorders.

One reason that damage of any kind to the hippocampus can lead to a wide variety of disease states is because of its anatomical location. It receives either direct or indirect input of information from virtually every sensory modality and from many secondary association areas. It also receives input from a number of subcortical regions involved in basic regulatory mechanisms that keep your body doing what it needs to do to survive. Thus the hippocampus lies at a critical position between evolutionarily older and newer brain regions, and serves to integrate much of this information.

Principal cells in the hippocampus are very sensitive to stress induced by either early life trauma or direct infusion of glucocorticoids. Biologist Bruce McEwen of The Rockefeller University has performed an extensive series of studies examining the effects of excessive glucocorticoid concentration on cells in the hippocampus. His work, and that of others, points to three very important forms of "structural plasticity" that occur in the hippocampus of a severely stressed animal.

The term "structural plasticity" is usually reserved for dramatic changes that involve the cytoskeletal structure of the cell. That is, the actual infrastructure of the cell is altered as a result of some experience. McEwen, along with Stanford biologist Robert Sapolsky (you'll remember him from Chapter 3) showed in the mid 1990s that adrenal steroids (glucocorticoids) cause a reversible remodeling of hippocampal cell dendrites[1]. Dendrites are complex, thread-like extensions that emanate from the cell body of a neuron. They are loaded with receptors that are specialized to bind neurotransmitters released from presynaptic cells. They are sort of the 'antenna' of the neuron, used to receive the many signals sent from presynaptic cells.

In one study, rats were given either 21 days of glucocorticoid treatment or 21 days of restraint stress (for several hours per day). Both manipulations resulted in significant remodeling of hippocampus cell dendrites as compared to control rats that did not experience either manipulation. When this discovery was first made, the authors used the phrase "neural atrophy" to describe what looked like a shriveling of the dendrites (illustrated in Figure 6-1). Indeed, either stress-related manipulation tends to reduce the complexity of the branching of cell dendrites and their size. Recent evidence, however, has shown that such changes are reversible within 7 to 10 days of the termination of the stressor – hence the term "remodeling".

This is quite intriguing and relevant to our discussions of mood disorder, since several studies have shown that prolonged, major depression is associated with a selective reduction in hippocampus volume that typically persists long after the depressive symptoms relinquish. Indeed, the most replicated findings regarding changes in the human brain that accompany mood disorder are decreases in regional volume, and the magnitude of the volume reduction is typically correlated with the severity of the depressive symptoms. Most often these reductions have been shown in the hippocampus and prefrontal cortex, although it is not clear at present if such reductions in brain volume are a cause or symptom of depression (like many other physiological changes that occur with depressive disorder).

One theory is that volume reductions of this kind may result from a process much like the dendritic remodeling seen in rats with excessive glucocorticoid circulation (Figure 6-1). An interesting study would be to determine the degree to which these brain regional volume reductions correlate with an overactive HPA axis in patients suffering from current mood disorder. There is data indicating that patients with post-traumatic stress disorder (PTSD) or Cushing's syndrome, the majority of which have exaggerated stress responses and excessive glucocorticoid secretion, show very similar patterns of volume reduction.

The neurochemistry of dendritic remodeling is a subject of intense research at the moment. This is because we think something very similar to this process may occur naturally with aging. Once the molecular/cellular mechanisms are understood, it may be possible to design drugs that slow the process and prolong function (such as learning and memory).

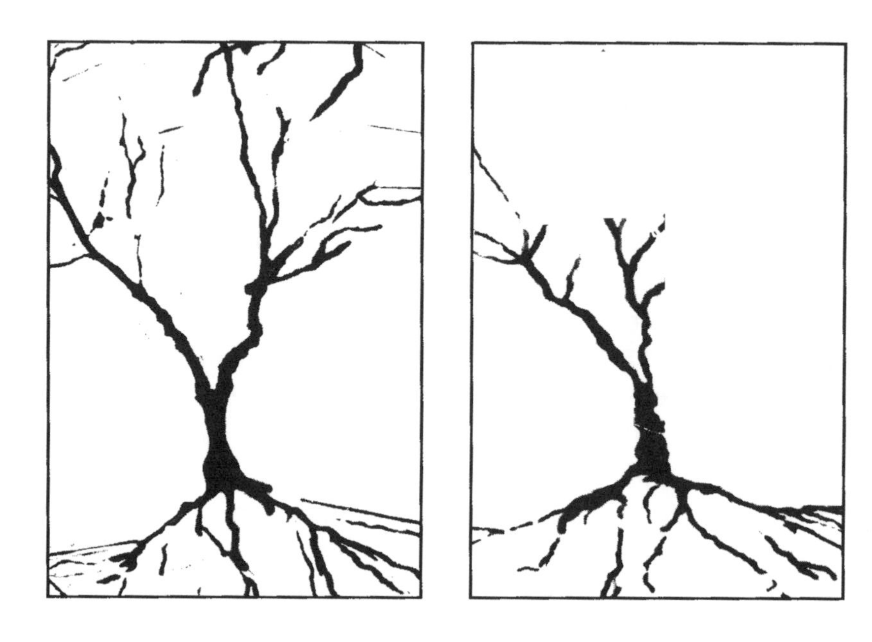

Figure 6-1 – A schematic illustration of the way a healthy neuron (left) can undergo dendritic remodeling induced by stress (right).

A critical step toward this goal came when it was discovered that dendritic remodeling can be prevented by the anti-epileptic drug, phenytoin (Dilantin). This finding suggests that excessive glutamate (an amino acid that works as a neurotransmitter throughout the brain) release from cells may be an essential ingredient for the remodeling. Phenytoin and many other anticonvulsants, prevent seizure activity by reducing the release of glutamate from nerve cells. When glutamate binds to a postsynaptic cell, the reaction typically makes the cell more excitable. If the cell becomes too excited, it will in turn release excessive glutamate that overexcites even more cells, and so on. Once a critical level of glutamate is surpassed, this process become toxic and neurons in that region begin to exhibit a state of decay. They often show signs of dendritic degeneration, and eventually the entire cell will die as a result of "excitotoxicity". This form of cell death via glutamate-induced excitotoxicity has been hypothesized to be a critical component in a number of neurodegenerative disorders including Alzheimer's to Parkinson's disease.

Consistent with the theory, study after study has now shown that glucocorticoids facilitate the release of glutamate in the hippocampus and prefrontal cortex. Moreover, stress-induced dendritic remodeling is blocked by the administration of adrenal steroid synthesis inhibitors such as cyanoketone, suggesting a pivotal role for glucocorticoids in the remodeling process.

Neurogenesis in Adults

When I was in graduate school the prevailing opinion in the classroom and textbooks, was that while it's common to see cellular regeneration and genesis in the peripheral tissues of adult animals (for instance when a wound heals), these processes do not occur in the central nervous system. The idea was that once you cross a developmental milestone, no additional neurons are generated. Once brain cells die – that's it, they are not replaced.

Of course, there are examples of cellular genesis and proliferation in developing animals, but the idea was that once you get past a critical period, no new nerve cells are generated. There was that one pesky study in the 1970s that *seemed* to show neurogenesis in the adult hippocampus. But nobody wanted to believe the data, and those who did wondered if the new cells were at all functional. In other words, do they become an active part of the brain's circuitry?

Although it has astonished many in the neuroscience community, there is now compelling evidence that new brain cells continue to be generated in adult animals! The growth and proliferation of new hippocampus neurons, has been demonstrated in a variety of adult mammals including humans. Even more amazing is the fact that we now have data indicating these newborn cells mature and become integrated into the hippocampal circuit and apparently contribute to the functioning of this brain region.

Neuroscientist Elizabeth Gould and her colleagues at Princeton University have performed a series of studies investigating the hormonal and experiential factors that regulate the production and survival of the newborn cells. Two positive regulators of adult neurogenesis are *estrogen levels* and *environmental complexity*[2].

The ovarian hormone estrogen facilitates the production of new hippocampus cells, while removal of circulating estrogen by ovariectomy results in a marked decrease in the proliferation of these cells. Thus, as you may have guessed, female animals (at least rats) produce more new hippocampus cells in adulthood than males, but many of them appear to degenerate so quickly that by several weeks after mitosis, these gender differences are negligible.

Environmental complexity and new learning have also been found to stimulate the production of hippocampus cells. Animals living in enriched environments with daily stimulation and contact with peers develop substantially more new cells than those animals maintained in standard laboratory colonies. Additionally, animals that are required to learn new tasks or exercise on a daily basis generate more hippocampus cells than their dull, couch-potato counterparts.

The possible therapeutic ramifications of these findings are still being grappled with. It's quite extraordinary (and perhaps premature) to think of the potential clinical utility of this phenomenon. Some of the challenges at present reside in our becoming more familiar with the negative regulators of new cell proliferation and survival, and learning how to avoid them.

Three of the most consistently found negative regulators that reduce neurogenesis and cell survival in adults include *social and environmental deprivation*, *stress*, and (you guessed it) *excessive adrenal steroid levels*. By deprivation, I mean the opposite of enrichment. This could be considered a lack of cognitive, motor, or social engagement on a daily basis.

Exposure to any number of stressful stimuli that typically elicit an increase in glucocorticoid secretion leads to a dramatic reduction in new cell proliferation. Incredibly, exposure to only a *single* stressful incident has been shown to produce a significant decrement in hippocampus cell proliferation and survival. These decrements are preventable by removing the adrenal gland (and thus decreasing glucocorticoid levels) or administering glucocorticoid synthesis inhibitors. Thus the mechanism by which stress shunts neurogenesis seems to be mediated by glucocorticoids.

Several questions emerge from these data in relation to clinical disturbances of mood:

1. Is it possible that depressed patients who exhibit reductions in hippocampus and prefrontal cortex volume are affected by a

glucocorticoid-induced reduction in neurogenesis?
2. Are these volume reductions a cause or effect of mood disorder?
3. How do changes in structural plasticity as seen by dendritic remodeling, cell death, and/or decreases in neurogenesis contribute to depressive symptoms?
4. If I'm depressed, should I run out and buy estrogen supplements?
5. …how about joining a gym, going back to school, or joining a social club?

I wish I could tell you we have the answers to these questions, but for now we have only theory, partial answers at best, and even more questions. There is accumulating evidence suggesting that depressives who show signs of hippocampus volume reduction do indeed also tend to have elevated glucocorticoid levels. But is this a cause or effect of the disorder? (I'm sure you're getting tired of that question by now.) We just went through data showing that glucocorticoids can produce dendritic remodeling, overt cell death, and decreases in neurogenesis. So what's the problem?

For starters, decreases in hippocampus cell numbers may *produce* an increase in glucocorticoid circulation because (as you'll recall from Chapter 3) they are involved in maintaining the proper levels of this hormone through a negative feedback mechanism. If you remove or damage the negative feedback mechanism (i.e. the hippocampus), glucocorticoid levels will rise unabated. So, once again we are faced with this insistent question – is this a cause or effect of depressive symptoms.

The most realistic answer is that we just don't know. Not yet anyway. More research is needed to determine: (1) if glucocorticoids play a causal role in hippocampus volume reduction; and (2) how changes in hippocampus circuitry may contribute to the symptoms associated with mood disorder. There is clearly evidence from both laboratory animal and human studies that some of the brain areas involved in the stress response are also involved in other functions that are typically affected during bouts of clinical depression. For instance, the locus coeruleus and raphe nuclei are involved in sleep cycle regulation; the hypothalamus is intimately involved in appetitive drives such as for food and sex; the amygdala and hippocampus both have major roles in learning, memory, and the regulation of mood, to name just a few. Thus a rather broad circuit is emerging in modern theories of mood and its disorders, and this circuit has considerable overlap of function with the stress response.

Even though we may not yet have a complete understanding of the causal sequences that are at work in this circuitry, we can still take advantage of the overlap in looking for effective treatments of mood disorder. As we've already discussed, these observations suggest that a new class of pharmacological agents that may have antidepressant properties are those substances that weaken a hyperactive HPA axis response to stress. Several drugs that work in this manner, by either disrupting CRF binding to receptors in the pituitary, or reducing glucocorticoid synthesis, are currently being tested in clinical trials for their efficacy as antidepressants. The results thus far are encouraging.

Just as exciting as the possibility for new antidepressant drug therapies that target the HPA axis and stress response, is the potential for behavioral therapies that target these systems. Behavioral treatments that reduce a hyperactivated HPA axis response to stress (or exaggerated basal levels of response) may have antidepressant properties. The link between one such behavioral action, reduction of the stress response, and the amelioration of depressive symptoms seems undeniable – namely exercise. Numerous studies have shown that daily exercise reduces mood disorder symptoms, often as affectively as antidepressant drugs. Human and laboratory animal studies have demonstrated that routine exercise reduces HPA axis signatures such as levels of circulating CRF and glucocorticoids, as well as physiological measures of the response such as heart rate and arterial blood pressure. Other experiments (some we have just discussed) have shown that exercise increases hippocampus neurogenesis and can prevent (and in some reports actually reverse) reductions in brain volume in stressed laboratory animals. A similar study has yet to be performed in humans with or without clinical depression.

Summary: The Stress-Diathesis Model of Depression

Early life trauma in the form of abuse, loss, or chronic stress often has long-term effects on the HPA axis, the autonomic nervous system, and the monoamine systems. The hyperactivation of the HPA axis and norepinephrine system has several important consequences.

Behaviorally, such persistent changes may lead to a number of symptoms associated with mood disorders such as sleep and appetite irregularities, anxiety, and alterations in emotional processing. At the anatomical level, persistent elevations in glucocorticoids induce changes in the physical structure of nerve cells located in the hippocampus and perhaps prefrontal cortex. High levels of glucocorticoids facilitate glutamate release from principal cells in these regions, and have been linked to several changes in structural plasticity, namely dendritic remodeling, cell death from excitotoxicity, and a decrease in the proliferation and survival of new cells (neurogenesis). These structural changes are thought to underlie the decreases in hippocampus and prefrontal cortex volume seen in patients with a history of mood disorder.

At the moment we do not have a very good understanding of the behavioral symptoms such structural changes may produce. However, several researchers have suggested the structural abnormalities may be a causal factor in the subsequent development of anxiety and mood disorders.

At the heart of this model (shown in Figure 6-2) is the premise that life experiences and genetic disposition interact in rather specific ways to regulate mood and can lead to affective disorders. Genetic variables contribute to this interaction in many ways, for instance, as a predisposition toward developing a hyperactive HPA axis response to stressors. The dependency on this type of interaction as a *causative* factor has led to the descriptive name, the "stress-diathesis model".

One of the leading proponents of the model, Charles Nemeroff of Emory University, suggested that mood disorder results in some individuals from an interaction between "vulnerability genes", "resistance genes", and adverse experiences encountered early in life[3]. He and others who developed the model acknowledge, however, that it may not apply to all patients suffering from affective disorders. The model has inherent limitations since not all depressives report early life trauma, hyperactive stress responses, or changes in CRF and cortisol production. Furthermore, not all depressives show the pattern of localized volume reductions to the hippocampus and prefrontal cortex that have been discussed in the last few pages.

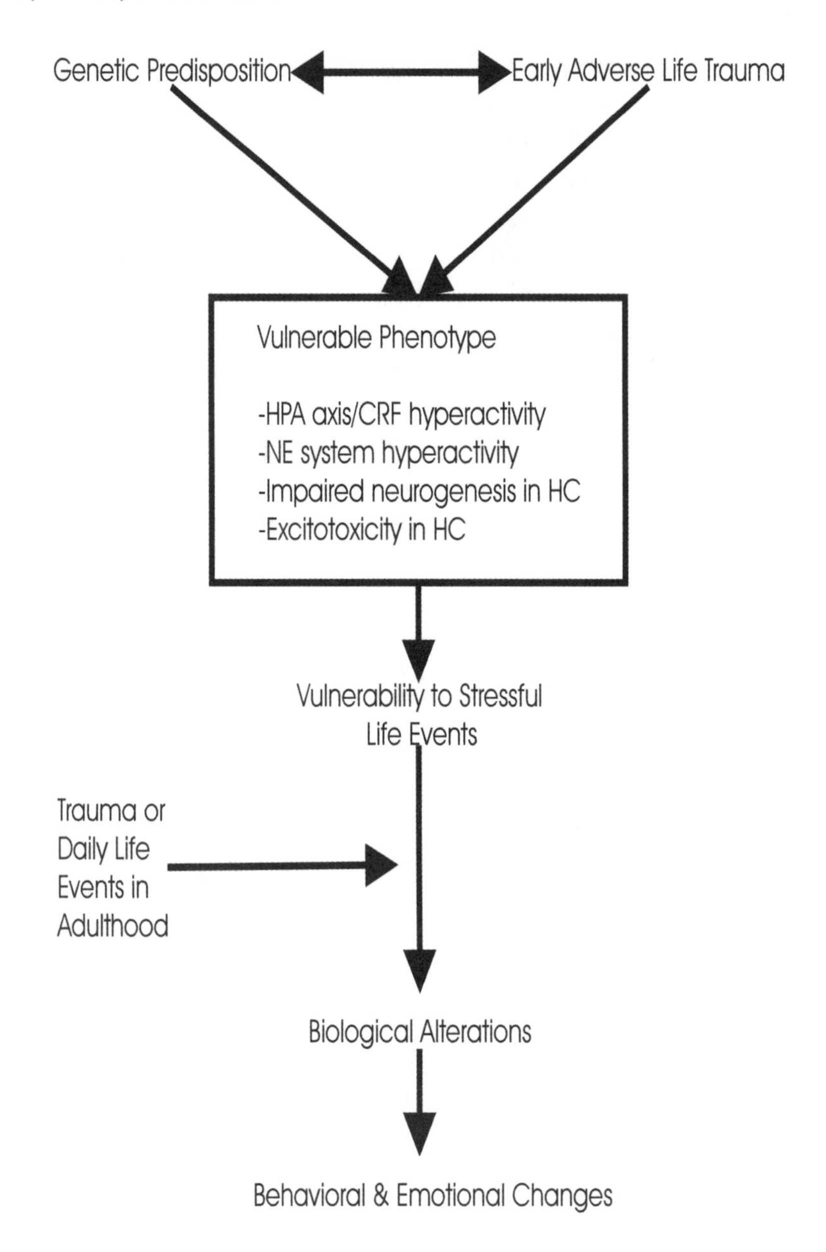

Figure 6-2 – A schematic illustration of the "stress-diathesis model" of depression. [Adapted from Update on Neurobiology of Depression, Psychiatry Online]

Similarly, many patients with mood disorders do not exhibit a negative response to the Dexamethasone Suppression Test, at one point thought to be a reliable physiological marker for mood disorders (see Chapter 2).

Does this mean that there are no reliable physiological markers for detecting anxiety and mood disorders? No, indeed these observations suggest there are perhaps *several biological routes* to major depression. It's also likely that different individuals will have greater pathology in one part of the circuit that we have been discussing than others. That is, one person may have problems that are predominantly relegated to the HPA axis, but eventually spread to the monoamine systems. Another individual may show the opposite pattern – an initial pathology in the monoamine systems that begins to creep into the HPA axis and endocrine system.

Chapter 7

†

Monoamine Theories of Mood: Revisited & Revised

There are many ways to look at how theory drives experimentation in the biomedical sciences. One school of thought tends to live by the credo that a theory should be "revised and revised until the data complies". Theorists from this school begin with a basic idea of how some process works, and keep adding bells and whistles to the theory until it can account for any new discrepant findings.

Another approach is found in the sentiment of Occam's Razor – which says all things being equal, the simplest theory is the best. This school tends toward parsimony. Good theory tends to lie between these two extremes – looking for the simplest explanation of a given phenomenon, yet acknowledging the fine details that contribute to the process.

Please allow me to apologize in advance for the seemingly bewildering array of revisions, extensions, and modifications to the basic monoamine theory of mood that was born from observations in the 1950s. (We discussed these historical events in the introduction.) You'll recall that through a series of serendipitous discoveries, a picture began to emerge suggesting that affective disorder symptoms are caused by a depletion of monoamines, particularly norepinephrine and serotonin. The evidence seemed pretty straightforward, drugs that reduced these neurotransmitters such as reserpine caused depression, while other drugs that had antidepressant effects were found to increase the levels of monoamines (e.g. tricyclics and MAO inhibitors – see Introduction).

One early question that arose from this theory concerned *which* particular monoamine transmitter was playing the key role in generating the symptoms. Tricyclics, reserpine, and MAO inhibitors affect several monoamines at the same time, especially serotonin and norepinephrine systems. So which is it?

In the last two decades, several key discoveries began to shed light on the functional roles of these systems. Through a mapping of pharmacology, neuroanatomy, and physiology, a picture is slowly (very slowly) emerging that helps describe how such a vast array of symptoms can arise to collectively produce what we refer to as disorders of mood. Let's start with the norepinephrine system.

The Norepinephrine System: Some Facts & Theory

It should be clear to readers by now that it's often fairly difficult to evaluate the function of an isolated transmitter system. We saw that each of the monoamine systems have a number of interactions with the stress response system and each other. Thus it is important to recognize that just because science makes discrete categories that are demarcated by protein and chemical structure, nature is not obliged to do so with respect to function. There is considerable overlap of function across the different monoamine systems. Nevertheless, we will point out some of the less controversial interpretations of experimental data from laboratory animals and humans that suggest where these systems can be delineated and how each may contribute to the symptoms observed in mood disorder.

As we discussed in Chapter 5, most of the cells that contain norepinephrine are located in a brain stem region called the locus coeruleus. When cells in this region become excited, they release norepinephrine into the neocortex, hypothalamus, thalamus, cerebellum, spinal cord, and a variety of other brain stem locations as shown in Figure 7-1. Through both animal and human studies, several functions have been found to depend on the norepinephrine system.

Typically these are discerned by comparing an animal that has had no manipulations with another that has had one class of receptors blocked (e.g. for only one subtype of serotonin receptors). Since synaptic transmission at these receptor locations will not occur as usual, any behavioral impairments that are observed in these animals compared to untreated controls are interpreted as being at least partially regulated by that transmitter system. First you show that this animal can perform a given function. Then if blocking a particular set of receptors impairs the function, one has evidence that the transmitter system in question is *necessary* for the function. This does not, however, mean it is *necessary and sufficient* for the function to occur. There could be other components involved in the ultimate expression of the behavior that go well beyond the transmitter system under scrutiny.

A second procedure for identifying the functional involvement of a transmitter system is to determine the kinds of behaviors/tasks that generally evoke activity in the system. What turns the system on? For norepinephrine, one may start to answer this question by determining which behaviors activate cells in the locus coeruleus and cause them to release transmitter.

Neurons in the locus coeruleus are most active when a novel stimulus or task is presented demanded of a subject - something new and exciting has to be happening. They are least active when the subject is just sitting around with no demands for vigilance. This effect has been replicated so many times and in assorted experimental contexts that most researchers feel the main function of the norepinephrine system is in supporting general arousal. In particular, this system is thought to be involved in a number of select functions that fall under the rubric of regulating attention, including learning and memory, pain sensitivity, anxiety, mood, and sleep-wake cycles.

Drawing on data from laboratory animal experiments and clinical studies in humans, a set of functional impairments or disturbances has been developed that may be predicted to occur with norepinephrine transmission deficits. These are shown in Table 7-1.

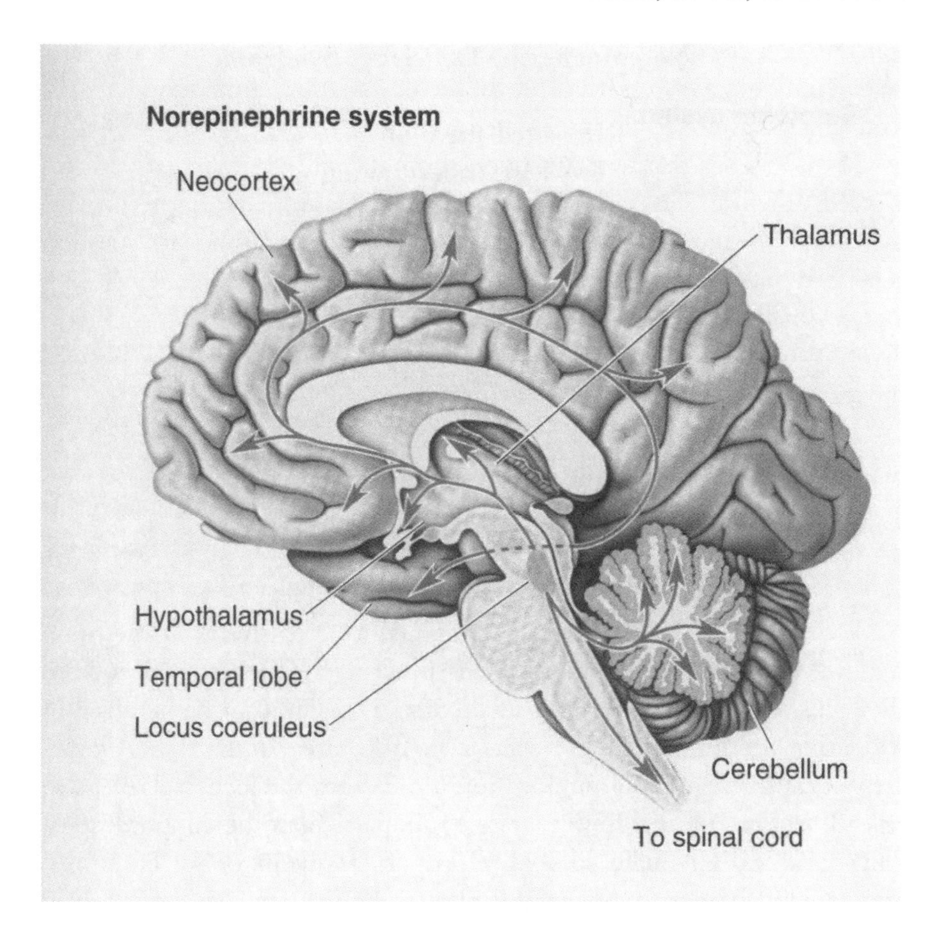

Figure 7-1 – The norepinephrine system. The majority of norepinephrine-containing cells are located in a small area of the brain stem known as the locus coeruleus. These cells connect to a number of target regions including the cerebral cortex, thalamus, hypothalamus, temporal lobe, brain stem, and spinal cord. [Figure adapted from Bear et al. (2001) *Neuroscience: Exploring the Brain*, Lippincott, Williams, and Wilkins]

Norepinephrine Deficiency Syndrome

- Impaired attention
- Problems concentrating
- Slowness of information processing
- Impaired working memory
- Depressed mood
- Psychomotor difficulties
- Fatigue

Table 7-1- Several hypothesized functional deficits accompanying norepinephrine deficiency.

There are several distinct norepinephrine pathways in the brain that are thought to mediate different functions. The connection from the locus coeruleus to the prefrontal cortex is believed to be involved in regulating attention, while connections to other areas of the frontal cortex are critical for regulating mood. Projections from the locus coeruleus to several regions of the limbic system, in particular the hippocampus, amygdala, and cingulate cortex, are important mediators of emotions, arousal, and memory, while connections to brain stem sites are involved in regulating blood pressure and heart rate.

There are also several *subtypes* of norepinephrine receptor, differentiated by their protein structure and function. Some norepinephrine receptors are located on the postsynaptic cell and receive the transmitter signal from the presynaptic cell in the fashion we have been discussing in earlier chapters. Other norepinephrine receptors are located on the presynaptic cell, very close to the site where the transmitter is released. This type of arrangement is called an "autoreceptor", since the receptor is on the same cell that is releasing the transmitter.

The function of autoreceptors is to regulate the amount of transmitter being released in a homeostatic-like manner. For instance, if too much transmitter is released into the synapse, some of it will drift across and bind to receptors on the postsynaptic cell, but extra transmitter will bind to the autoreceptor. When this happens, the presynaptic cell knows it's releasing too much (since extra is drifting back to the release site), and it reduces the synthesis and further release of the substance.

The opposite effect also occurs - synthesis and release of transmitter can sometimes increase until extra molecules of the substance drift back and bind to the autoreceptor. In this fashion, autoreceptors facilitate the ability of the presynaptic cell to self-regulate its release of transmitter. As you will see in the next sections, because there are so many receptor subtypes and locations, the relationship between transmitter levels and mood disorder symptoms is not at all straightforward. Before we jump ahead, however, let's consider some basics about the serotonin system.

The Serotonin System: Some Facts & Theory

The origin of the serotonin system is a series of brain stem cell groups called the raphe nuclei. Neurons in the raphe nuclei connect, like the norepinephrine system, to a wide range of target sites including the neocortex, thalamus, hypothalamus, basal ganglia, several limbic and brain stem regions, and the spinal cord (see Figure 7-2). Serotonin-containing cells in the raphe nuclei become activated by a number of different conditions, thus it is thought that the system contributes to several functions.

Connections from the raphe nuclei to the frontal cortex and limbic structures, including the hippocampus and amygdala, have been shown in laboratory animals and humans to be involved in the regulation of mood. Pathology in these pathways has also been reported to contribute to anxiety and panic disorders. Connections from the raphe nuclei to motor areas of the basal ganglia (a subcortical area) that are normally involved in regulating movement are suspected to play a role in the development of compulsive behaviors. Finally, a large projection to the hypothalamus is known to regulate appetite and eating activities, while projections to brain stem sites are responsible for maintaining proper sleep-wake cycles.

Considering all these pathways and their putative functions, one can piece together a set of dysfunctions that may appear as a result of serotonin depletion. As is illustrated in Table 7-2, many of these disturbances are components commonly observed in mood disorders.

Serotonin Deficiency Syndrome

- Anxiety
- Depressed mood
- Panic
- Changes in sleep-wake cycles
- Obsessions and compulsions
- Phobia
- Changes in appetite and eating behavior

Table 7-2- Several hypothesized functional deficits accompanying serotonin deficiency.

Like the norepinephrine system, serotonin receptors are found on both the postsynaptic and presynaptic (autoreceptors) sides. An interesting twist is that serotonin-containing cells not only have autoreceptors that bind serotonin, but also have receptors for norepinephrine that modulate the release of serotonin. Thus norepinephrine levels in the raphe nuclei can directly affect the subsequent levels of serotonin at target brain areas[1]. The situation is fairly complicated – norepinephrine can either facilitate or hinder the release of serotonin depending on which subtype of receptor is involved.

Let's now examine some general properties that are common to both of these systems and consider extensions of the classical monoamine theory of depression that account for earlier discrepancies.

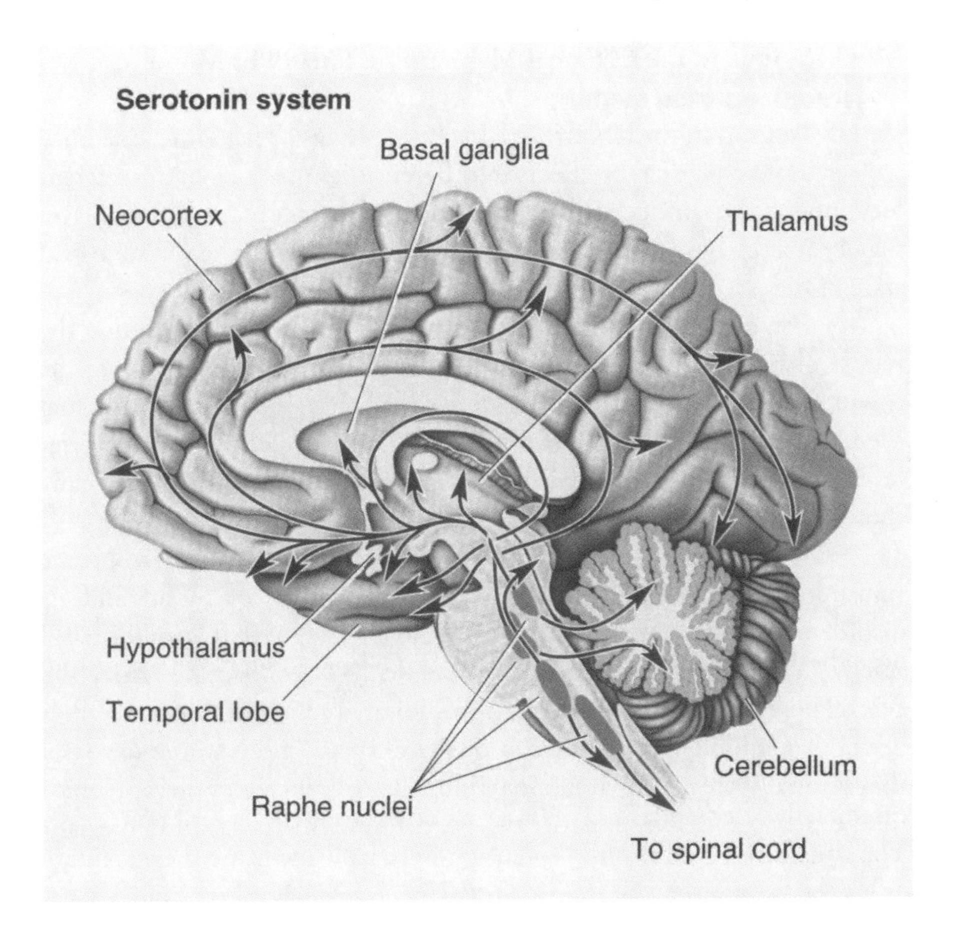

Figure 7-2 – The serotonin system. The majority of serotonin-containing cells are located in an area of the brain stem called the raphe nuclei. These cells project into a number of target regions including the cerebral cortex, basal ganglia, thalamus, hypothalamus, temporal lobe, brain stem, and spinal cord. [Figure adapted from Bear et al. (2001) *Neuroscience: Exploring the Brain*, Lippincott, Williams, and Wilkins]

Modifications to the Classical Monoamine Theory of Mood

Neurotransmitters are tricky little things. The same one may have completely different effects in different brain regions or at different locations on the same cell. The ultimate action that is observed depends on which subtype of receptor the transmitter binds to, and the physiological state of the cell.

For example, norepinephrine can either enhance or shunt the release of serotonin from the same cell, depending on the subtype of the receptor that is activated (alpha 1 or alpha 2). It's perhaps not surprising then, that a simple straightforward relationship between transmitter levels and emotional tone has proven to be more elusive than originally thought.

The earliest (and perhaps still the best) evidence for specific transmitter systems being involved in the etiology of mood and its disorders comes from findings in the 1950s and 1960s that were reviewed briefly in the introduction. Drugs that deplete monoamines from the central nervous system produced depression in a subset of subjects. Other drugs that inhibit the production of the enzyme (monoamine oxydase) that normally breaks down monoamines and thus increases their presence in the synapse between two cells were found to have antidepressant properties.

Likewise, agents that block the reuptake of monoamines back into the presynaptic cell (tricyclics) were found to have antidepressant actions. So the picture seemed pretty clear – manipulations that reduce brain monoamines produce depression, while manipulations that increase monoamines alleviate the symptoms. If only it were that simple.

There are several problems with such an interpretation, many of which were known decades ago. First, only a small percentage (approximately 20%) of individuals who were treated with monoamine depleting drugs, such as reserpine, exhibited depressive symptoms. Furthermore, most of the patients that exhibited reserpine-induced depression had a previous history of the ailment.

In a similar vein only about 50-60% of patients suffering from clinical depression respond to treatment with either tricyclics or MAO inhibitors. By comparison, approximately 30% respond to treatment with a placebo (sugar pill). (Also, other drugs that increase brain monoamine levels such as cocaine and amphetamines are not very effective antidepressants.) Indeed, these are all significant numbers, but are they convincing? A sad fact of science is that, like any other industry it has a momentum that is in part driven by money and marketing. In his book, *Blaming the Brain*, Elliot Valenstein examines the many ways the drug industry, in its desire to manage cost/risk ratios, has focused on the monoamine hypothesis to guide research and development of antidepressants at the expense of other possible avenues. The total market value of antidepressant drug sales worldwide for the year 2000 was estimated to be in excess of 6.5 billion dollars. "Considering how much money is involved, it is no surprise to find that economic factors play a powerful role in all stages of drug development, research, clinical trials, and marketing."

Today, there are still so many unanswered questions about the monoamine theory of mood. For example, it is still unclear *which* transmitter system plays the pivotal role in the etiology of depression. This is because tricyclics and MAO inhibitors affect all monoamines, although to different degrees. Earlier accounts of the monoamine theory suggested that norepinephrine depletion was the chief cause of depression. Then the SSRIs were developed, the most famous example being flouxetine (Prozac), and found to be just as effective in alleviating depressive symptoms, but with less side effects than the tricyclics and MAO inhibitors. This encouraged a shift in thinking toward the notion that mood disorder is principally caused by a reduction in serotonin transmission. As this book goes to press, clinical trials are being completed on the first series of selective norepinephrine reuptake inhibitors (SNRIs), such as Reboxetine. The theoretical pendulum will no doubt continue to swing.

Another, even more troubling question, concerns the timescales of antidepressant actions. Tricyclics, MAO inhibitors, SSRIs, and SNRIs alter monoamine levels in the brain almost immediately, but it generally takes 3-4 weeks for these substances to alleviate depressive symptoms. Why should antidepressants take so long to work if the mechanism that is supposed to be causing the symptoms is rectified within minutes? There is a simple answer – it's probably not the correct mechanism. With all the data that has now been collected in the last fifty years in a variety of laboratory and clinical settings, it has become more and more difficult to accept the suggestion that mood disorder symptoms are caused directly by monoamine depletion. So what are some of the alternatives?

Monoamine Receptor Hypothesis

One alternative theory that has become very popular in the last decade is that *mood disorders are actually caused by an excess of monoamines*. Okay, here's the logic. As we've noted, typical antidepressant drugs temporarily increase monoamine transmission. But remember a while back we talked about how nature loves to be in equilibrium. We said that biological mechanisms are driven toward allostasis – a kind of dynamic equilibrium. If an external agent causes a change in the brain, compensatory reactions begin to occur to return the system to its proper operating range.

If excessive monoamine levels cause depression, initial treatment with typical antidepressants that increase these levels even more should temporarily exacerbate the symptoms. Indeed many clinicians insist this is what occurs. However, once the receptors for the monoamines detect chronic increases in their levels, they will begin to down-regulate. You'll recall this is a way the receptor compensates for excessive transmitter in the synapse. This can occur by a reduction in the number of monoamine receptors or in existing receptors being less responsive to the transmitter. Interestingly, this process of down-regulation of receptors has been estimated to take approximately 3-4 weeks to occur – about the same time it takes antidepressants to alleviate depressive symptoms. As the compensatory changes occur in the receptors, less monoamine *transmission* takes place – the transmitter is there, but the receptor efficacy is not.

A critical assumption is that antidepressant treatment results in an over-compensation at the receptor level. Imagine you start with a baseline level of transmission and then double it with antidepressant treatment. If the compensatory reaction of the receptors is to decrease their efficacy by half, these two actions cancel each other out. However, if the compensatory reaction of the receptors is larger than the initial increase in monoamine levels, the net result is a *decrease* in monoamine transmission. In this scenario, symptoms of depression are alleviated once monoamine transmission decreases back to a normal physiological range, and antidepressants actually work by inducing these long-term compensatory reactions at the receptor level. This theory does address the timescale issue, but there are still many unanswered questions.

For instance:

1. What fosters the initial increase in monoamine transmission leading to the depression?
2. If the brain produces compensatory reactions at the receptor level in response to antidepressant treatments, why can't it do so with the initial increase that originally caused the mood disorder?
3. And the big one – How does an increase in monoamine transmission produce mood disorder symptoms?

These are just a few of the questions that need to be addressed to make this theory stand on firmer ground. I believe the severity of our current state of confusion is illustrated well by the fact that modern theorists can't even agree on whether clinical depression is caused by an increase or decrease in monoamine transmission. And at present it is certainly unclear what the mechanism of treatment really is. Do the receptors really over-compensate? We do not yet know the answer.

A second alternative hypothesis to the classical monoamine theory is that mood disorder is indeed caused by a deficiency in monoamine transmission, but in a more complicated way than was originally thought. We talked about autoreceptors in earlier sections and I just know you've been waiting to hear more about them.

Imagine antidepressants are given to a patient. As was the case in the last scenario, this will cause a temporary increase in monoamine transmission, followed by compensatory changes at the receptor level. However, in this scenario, both postsynaptic and presynaptic (autoreceptors) become down regulated. What will be the result of such a process?

Well, it depends. Let's suppose that the postsynaptic receptors down-regulate to a greater extent than the presynaptic autoreceptors. In this case postsynaptic receptor down-regulation will result in less receptor binding and as a whole less monoamine transmission. If the autoreceptors are down regulated as well, this will lead to them being less sensitive to excessive transmitter in the synapse and therefore the presynaptic cell will tend to *release more transmitter*. But who cares how much transmitter is released if transmission is limited by a significant postsynaptic down-regulation. In other words the post-synaptic side becomes the limiting factor. You can have an overabundance of transmitter, but if there are fewer receptors for it to bind to on the postsynaptic side or if the receptors are less sensitive to the transmitter, monoamine transmission will still decrease.

Now suppose the opposite happens, that there is greater down-regulation of the presynaptic autoreceptors than of the postsynaptic receptors. What happens now? The prediction is that the autoreceptors become less sensitive, so the cell (in response to antidepressant treatment) releases more transmitter. Since the postsynaptic receptors have also been down regulated, they will indeed be less sensitive. *However, the assumption is that greater down regulation occurs on the presynaptic side, consequently the increase in transmitter release over-compensates for the slight reduction in postsynaptic receptor sensitivity.* In this scenario antidepressant treatments alleviate depressive symptoms by reducing the efficacy of autoreceptors, resulting in a facilitation of monoamine release and transmission. Here the limiting factor is assumed to take place on the presynaptic side.

Whew – did you get all that? Even I'm confused now. There are still plenty of questions. For instance, once the down-regulation of autoreceptors leads to an increase in transmitter release, is it not possible that this will eventually result in a down-regulation on the postsynaptic side? If so, then we're back where we started – a deficit in monoamine transmission. I'm sure by now you are beginning not only to appreciate the problems with the classical monoamine hypothesis of mood and its disorders, but also why it is so difficult to develop a reasonable extension that is not riddled with its own maddening difficulties.

There is now evidence indicating an increase in the number of serotonin receptors in the prefrontal cortex of suicide victims with a history of depression. The most parsimonious explanation for this would seem to be a reduction in serotonin transmission (at least to this region). However, to date evidence is lacking for either a monoamine transmitter or receptor abnormality that is consistently found in the majority of patients suffering from mood disturbances.

Monoamine Hypothesis and Gene Expression

Although there is still quite a bit of confusion and controversy surrounding the monoamine transmitter and monoamine receptor abnormality theories of mood disorder, a recent extension of these has been made that is concerned with post-receptor malfunctions.

Once transmitter binds to the receptor of a cell, a variety of different effects occur. Some signal second messenger proteins that make the postsynaptic cell more or less excitable. Other effects that have been observed include a direct modulation of gene expression patterns (see Chapter 1).

For instance, an excess of monoamine transmitter causes a repression of the gene that normally codes the construction of that particular class of receptors. This is the mechanism by which a down-regulation (through a decrease in receptor numbers) takes place. If the opposite happens, too little transmitter is released into the synapse, a series of reactions occur that eventually lead to an increase in the expression of genes that code for the construction of the complementary receptor (this is up-regulation).

The notion that perhaps disorders of mood involve an alteration in normal gene expression patterns was raised because even though evidence is inconsistent in terms of changes to monoamine transmitter levels or receptor numbers, these systems do indeed seem to work improperly in depressives.

So what are we left with? Well, the basic argument is that the regulatory mechanism is in the next step - once the transmitter binds to the receptor. People suffering from mood disorders may have a malfunction in the steps that occur between receptor binding of transmitter and the expression of genes leading to the construction of particular proteins. This hypothesis is made even more attractive by recent data showing that stress hormones can also affect the expression and repression of these same genes! This is currently one of the hottest research areas in the biology of mood and anxiety disorders. Let's look at an example.

Brain-derived neurotrophic factor and mood

Attempts are being made to identify the gene transcription factors that may malfunction in patients suffering from these disorders, and identify the target proteins that are affected. One exciting candidate mechanism is the gene induced by monoamine receptor binding that normally controls the production of a substance known as brain-derived neurotrophic factor (BDNF). BDNF is a trophic (or growth) factor that is very important for nerve cell survival. In the absence of BDNF a neuron may go into a state of programmed cell death, where it commits suicide by expressing certain substances that actively kill it. Thus BDNF plays an absolutely critical role in the survival and vitality of existing neurons and newborn cells.

When excessive stress hormones are released, a series of poorly understood reactions occurs that ultimately results in a repression of the gene for BDNF. Consequently, stress leads to a reduction of BDNF in the central nervous system. This is yet another important example of the way the monoamine transmitter systems interact with the stress response.

Reducing levels of BDNF in laboratory animals has been shown to result in cell loss in the hippocampus, and this effect has been posited to be a factor contributing to the observed volume reduction in the hippocampus of depressed patients. Several investigators have also suggested this mechanism may serve to increase the likelihood of recurrent episodes with greater exposure to the disorder.

It is well documented that individuals who have experienced mood disorder episodes in the past are far more likely to become depressed again. In fact the more times you have experienced a clinical episode in the past, the greater the likelihood of experiencing one in the future. Put in the language of gene expression, one can imagine a scenario by which each episode leads to a greater susceptibility to cell loss. A reduction in the number of existing hippocampus cells coupled with decreases in gene expression for BDNF can lead to even greater cell loss (and perhaps less responsiveness to treatment with repeated episodes).

This process is illustrated in Figure 7-3. Normally, monoamine transmitter binding to a receptor induces the expression of certain genes, some that code for the construction of other receptors, others that control the production of BDNF (and undoubtedly other trophic factors). The argument is that the transduction of monoamine receptor binding into gene expression is faulty in people suffering from mood disorder. Antidepressants, then, regardless of their initial actions on monoamine levels or receptor sensitivity, are thought to ultimately increase gene expression patterns for critical trophic factors like BDNF. It's likely that this, in turn, reduces hippocampus (and perhaps prefrontal cortex) cell loss and alleviates many symptoms of depression.

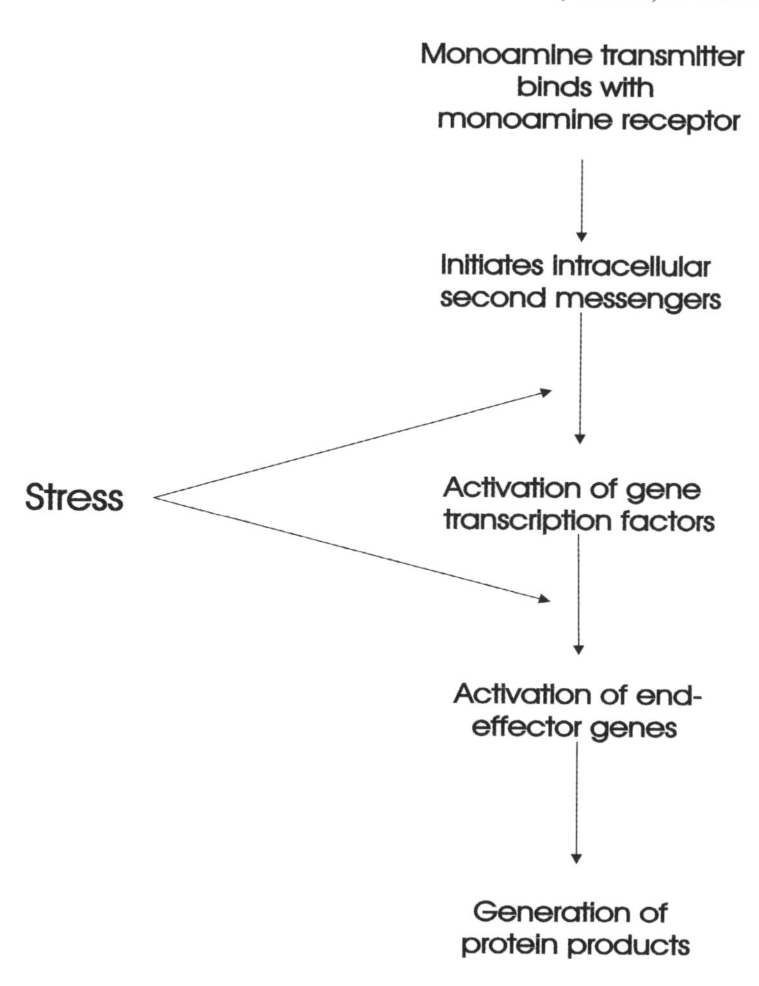

Figure 7-3 – Several effects may occur as a result of monoamine transmitters binding to a monoamine receptor. One effect is the initiation of a complex sequence of reactions that ultimately alters gene expression and the generation of protein products such as BDNF and postsynaptic receptors. Second messengers induce the activation of gene transcription factors that, in turn, regulate the expression of end-effector genes. Malfunctions may occur at any one of the steps in this sequence to produce abnormalities in gene expression and/or repression. For instance, stress is known to disrupt several steps in this process. In particular, stress

has been shown to repress the expression of the gene for BDNF, a trophic factor that is critical for the survival of existing neurons. Significant decreases in BDNF result in hippocampus cell loss that has been suggested to be a contributing factor of the volume reduction seen in many depressives. This process illustrates another important interaction between the monoamine systems and the stress response.

Interestingly, the sequence of events that contribute to stable changes in gene expression patterns in the hippocampus requires a period of time very close to that typically needed for antidepressant treatment to begin alleviating symptoms. A related footnote to this discussion is that there is now preliminary evidence that some antidepressants, particularly the tricyclics, can enter the nucleus of a neuron and bind directly to its DNA. Thus some of the therapeutic effects mediated by tricyclics may be due to their ability to regulate gene transcription, independent of their inihibition of monoamine reuptake.

The implications of this theory for treatment possibilities of psychiatric illness in general are enormous. One problem in developing effective treatments for brain disorders is in targeting the specific receptors in only the selective neural areas that exhibit pathology related to the ailment. A drug that binds to an alpha 1 norepinephrine receptor in the hippocampus will also do so in every other region of the brain and body. This could result in severe side effects associated with the treatment. It may also produce effects that actually offset the efficacy of the treatment.

However, much greater specificity may be possible if drugs can be developed that target gene transcription factors that uniquely regulate a single protein in a spatially restricted region of the brain. The development of such a drug, will not only reduce potential side effects, but also reduce the time it typically takes an antidepressant to begin alleviating symptoms since the actual transcription factor is targeted rather than a component much earlier in the sequence (e.g. monoamine levels).

This is similar in many respects to the "antisense technology" approach currently being used to treat other medical conditions such as cystic fibrosis. Once a gene has been related in some way to a disease process, and its DNA sequence is derived, probes can be developed that use the exact sequence of mRNA that complements the gene strand (see Box 2-1). An antisense probe can then be directed against the mRNA sequence, preventing the translation of the protein coded for by the gene. In this manner, even if early gene transcription factors are activated, the production of the final protein product will be prevented. Charles Nemeroff and his colleagues at Emory University have recently demonstrated a reduction in pituitary CRF transmission in the rat after administration of a CRF antisense probe. This suggests that available antisense probes for CRF may have utility in shunting a hyperactive HPA axis response to stress.

Other groups have since shown that infusion of CRF receptor antisense probes directly into the amygdala reduces anxiety-related behaviors in chronically stressed rats. These experiments are incredibly exciting and demonstrate the utility of the approach. However, additional studies are required to replicate these results in primates.

Chapter 8

†

Substance P and the Neurokinins

Recently, quite a bit of attention has been paid to a relatively new class of transmitters called neurokinins and their possible role in regulating mood. This theory has developed rather quickly in the last few years after serendipitous findings that an antagonist of one particular neurokinin, called substance P, may have antidepressant properties.

A provocative study published in 1998 in the prestigious journal, *Science*, reported antidepressant activity of a new neurokinin-1 receptor (the receptor type to which substance P binds) antagonist to be comparable to that of a leading SSRI, in a placebo-controlled, double-blind trial[1]. These observations have created a frenzy in research circles as scientists and pharmaceutical firms are now scrambling to replicate the results and design even more potent neurokinin-1 receptor blockers.

Preclinical Studies in Laboratory Animals

Substance P was first discovered in the 1930s, after being extracted from horse intestine and brain tissue. After further investigation, it was found that the extract caused fast smooth muscle contractions geared toward placing an animal in a state of "preparedness", hence the name, "substance P".

Traditionally, substance P was thought to be involved in the pain response because it is released from cells in the peripheral nervous system in response to inflammation. This demonstration led to studies designed to ultimately develop antagonists of the transmitter in the hopes that they will have analgesic effects. Although neurokinin-1 receptor antagonists have not shown considerable promise as analgesics, they have been found by some researchers to have a positive effect on mood.

Anatomical location of Substance P pathways

Among the family of neurokinins, substance P is by far the most concentrated in the central nervous system, and is found in the spinal cord, substantia nigra and striatum (two motor areas implicated in Parkinson's disease), hypothalamus, locus coeruleus, raphe nuclei, and several limbic structures including the amygdala and hippocampus. Autoradiography studies, which are designed to localize receptors, have shown the highest density of neurokinin-1 receptors overlap with the brain structures where substance P is most often found, particularly in the amygdala, hippocampus, hypothalamus, and locus coeruleus. In short, there is evidence that the distribution of substance P and the neurokinin-1 receptor overlaps to a significant degree with some brain pathways implicated in mood and anxiety disorders.

Many of these pathways are shared with the monoamine system. In fact there is evidence in both rodents and primates that substance P and serotonin are co-released from the same cell in some brain areas such as the raphe nuclei.

Another piece of evidence suggesting an interaction between these two systems is that injections of substance P into the cerebral spinal fluid of an animal produces increased levels of circulating catecholamines (norepinephrine, epinephrine, and dopamine). Consistent with this finding, infusion of substance P directly into the locus coeruelus increases cell firing in this region and presumably the release of norepinephrine into target brain sites. However, the evidence regarding the regulatory role of substance P on locus coeruleus and subsequent norepinephrine release is far from consistent. For example, some studies have demonstrated a reduction in locus coeruleus activity following administration of neurokinin-1 antagonists (which would be consistent with the above findings), but others have shown actual increases in cell firing.

How does Substance P influence behavior?

The most consistent result in the substance P literature is the effect of the compound on behavior. Administration of substance P produces aversion and anxiety-like behavior in laboratory animals. Similar results have been obtained using neurokinin-1 agonists (substances that bind to neurokinin-1 receptors and mimic the actions of substance P).

A behavioral test commonly used as an index of anxiety in rodents is called the "elevated plus-arm maze". In this task, a plus-arm shaped maze (shown in Figure 8-1) is positioned 3-4 feet above the floor. Two of the maze arms have sidewalls protecting the animal from falling, while the other two arms are unprotected. Rats are usually very inquisitive animals and are eager to explore novel environments. However, when placed in the elevated plus-arm maze they typically spend more time in the protected than unprotected arms. This gives the experimenter a baseline level of anxiety-like behavior by which to compare with various drug manipulations.

When drugs that are anxiety-reducing (anxiolytic) in humans are given to rats before this task, they spend about equal amounts of time in both protected and unprotected arms. In other words, they lose the anxiety associated with the fear of falling from an unprotected arm. Drugs that induce anxiety (anxiogenic) in humans result in rats spending even more time (compared to the baseline condition) in the protected arms rather than the unprotected arms (they become anxious).

Laboratory animals given substance P typically become very anxious and spend far less time in the unprotected arms of the elevated plus-arm maze, compared with control animals. These findings have been reproduced using other behavioral assays of anxiety, as well as physiological measures such as heart rate.

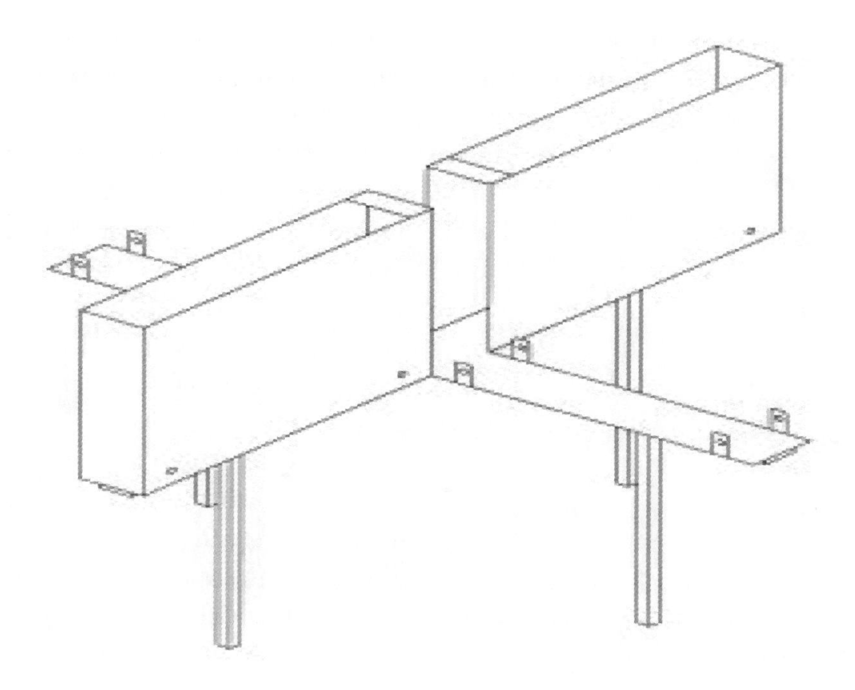

Figure 8-1 – A typical elevated plus-arm maze used to measure anxiety in rodents. Two arms (running southwest to northeast) of the maze are enclosed with a raised wall on both sides, protecting the rat from falling. Two additional arms (running northwest to southeast) have no walls, and therefore offer no protection from falling. Normal rats will typically spend more time in the protected arms than in the unprotected arms, providing a baseline measure of anxiety-like behavior. Anxiolytic drugs (which alleviate anxiety in humans) have been shown to shift this baseline so that animals spend approximately equal time in each of the arms. This suggests that animals become less fearful of the unprotected arms of the maze. Contrasting this, manipulations that are anxiogenic (anxiety producing in humans) shift the baseline in the other direction so that the animals spend even less time in the unprotected arms than was observed under baseline conditions.

Studies are now being performed that inject substance P directly into select brain regions so that the circuit controlling the substance P-induced anxiety can be determined. Our laboratory was among the first to demonstrate that substance P, infused directly into either the amygdala or hippocampus, produces anxiety-like behavior in rats. And very exciting (at least to us) is the fact that pretreatment of these rats with the SSRI flouxetine (Prozac), blocked substance P-induced anxiety. Other laboratories have now shown similar results in additional brain regions such as the hypothalamus. These experiments currently await replication in primates.

Experiments like these suggest that substance P and its activation of neurokinin-1 receptors may be modulated by stress. Indeed, there have now been several studies demonstrating that substance P-containing cells are sensitive to environmental stressors. Consequently, as would be expected, substance P levels change in different brain areas in response to stress. However, a consistent picture of changes has not been shown across brain regions.

Some regions like the hypothalamus and hippocampus have reductions in substance P concentrations following acute physical stress (footshock), while showing an increase in concentrations following psychosocial stress such as brief isolation or maternal separation. More studies are needed before we will have a reasonable understanding of how substance P regulates anxiety, and how stress influences substance P levels in the brain.

Contrasting results have also been reported with respect to the effects of antidepressant treatments on substance P concentrations. However, it is worth noting that these experiments vary considerably in their logistics, which may explain some of the inconsistencies. Most have shown that brief exposure to antidepressant treatments (14-days or less) has little or no effect on substance P levels in a variety of brain regions including the amygdala, hippocampus, hypothalamus, and cerebral cortex.

Other studies demonstrated a modest reduction in substance P concentrations in the amygdala following a 40-day antidepressant treatment plan. Consequently, although antidepressant treatment has been shown to block substance P-induced anxiety in laboratory animals, it is unlikely that such prevention is mediated by a direct interaction of antidepressants with substance P synthesis, release, or binding to neurokinin-1 receptors.

Thus far a rather cloudy picture has emerged. There is evidence that stress may cause increases in substance P concentrations in certain brain areas. There is also evidence that infusion of substance P into brain regions thought to be involved in mood and anxiety disorders results in anxiety-like behavior and that such responses are blocked by pretreatment with antidepressants. However, it is not clear that antidepressant treatments have any effect on substance P levels themselves. One explanation for this pattern of results may be that substance P and antidepressant agents have no direct interaction, but affect the same neural systems further downstream. If this is true, antidepressants and substance P antagonists may indeed share treatment efficacy in reducing symptoms associated with depression and perhaps anxiety disorders, but probably do so using entirely different mechanisms.

Earlier we mentioned one animal model of depression and anxiety, namely maternal separation. Even brief separation from the primary caregiver results in dramatic increases in HPA axis response to stress as adults, and can lead to additional behavioral changes in appetite, sleep-wake cycles, and social interactions with peers that are believed to closely resemble mood disorder in humans. At present, there are relatively few studies using this model to explore the efficacy of neurokinin-1 receptor antagonists as possible treatments for depressive and anxiety disorder induced by early life stress. Two studies have shown that animals pretreated with L733060, a neurokinin-1 receptor antagonist, are more resistant to maternal separation induced changes to the stress response system as adults as compared to untreated animals. This suggests that neurokinin-1 receptor antagonists may have some protective properties in reducing the long-term effects of early life stress on the subsequent development of depressive and anxiety disorders in adulthood. This is all well and good, but what about the human literature?

Clinical Studies of Substance P and Mood

Most clinical studies of substance P and its role in regulating mood have involved taking cerebral spinal fluid measures (using a lumbar puncture – also known as a spinal tap) of substance P and associated metabolites from depressed patients and comparing these levels with unaffected individuals that are matched for age, sex, and as many other variables as possible.

At least three independent studies have now shown that depressives have an increased level of substance P concentrations in their cerebral spinal fluid as compared to matched control subjects. One study found a four-fold elevation in substance P concentrations in depressed patients compared with matched control subjects, however, a subsequent study failed to find significant differences.

Postmortem analyses that compares brain tissue from depressives with matched controls after death (most subject are hesitant to volunteer brain tissue while still alive) has not revealed consistent differences in neurokinin-1 receptor numbers, suggesting little change between the two groups in substance P concentrations. So the jury is still out on these data.

One study that, as mentioned above, created quite a stir recently found that a neurokinin-1 receptor antagonist called MK-869 had approximately the same efficacy in alleviating depressive symptoms as an SSRI, but with less side effects. Scientists working at Merck Pharmaceuticals developed the compound that has been since shown to have a high affinity and selectivity for human neurokinin-1 receptors. Importantly, the compound has virtually no ability to bind to serotonin, norepinephrine, or dopamine transporter proteins, indicating that it does not alter monoamine reuptake properties. Moreover, it has no affinity for any of the monoamine oxydases, consequently it does not influence monoamine levels via this mechanism.

In a double-blind study of 213 depressed patients randomly assigned to treatment with placebo, the SSRI paroxetine, or MK-869, clinical improvement of symptoms was virtually identical in the latter two groups across the six-week study. Both of these groups showed a significant reduction in depressive symptoms compared to the group that received the placebo.

Also of significance, the MK-869 group showed fewer side effects than the SSRI group. In particular, they exhibited a decrease in the reported incidences of sexual dysfunction (3% vs. 26%) compared to the group being treated with paroxetine. This is good news since many patients report early termination of treatment (and subsequent relapse into depressive symptoms) due to this specific side effect. No noticeable differences were observed between the MK-869 and SSRI groups with regard to the time period required for alleviation of symptoms.

As interesting as this study is, depressed readers should not yet jump with delight. A single study does not establish unequivocally the effectiveness of a new treatment compound. Indeed, in a second trial, MK-869 exhibited approximately the same antidepressant efficacy as a placebo[2]. So this is where I again use the tired old cliché, "more studies are needed" to determine if neurokinin-1 receptor antagonists represent a new (and highly novel) class of antidepressants.

An even more potent neurokinin-1 receptor antagonist has recently been developed by Merck and introduced into clinical trials. The first trials should be completed by 2002. Based on the preclinical studies in laboratory animals and the few human studies performed to date, it is unclear how substance P pathways might contribute to mood disorder. Yet, the clinical trials performed by Merck (the first ever of a neurokinin-1 receptor antagonist) are certainly encouraging and will hopefully promote more research on the efficacy of these compounds in treating affective and anxiety disorders. Indeed, neurokinin-1 receptor antagonists may represent a radical new approach to the treatment of depression.

Chapter 9

†

The Immune System and Mood

Not very long ago, perhaps less than two decades, it was believed by most scientists that the brain and immune system had very little to say to each other. Your immune system regulates the production of antibodies that target infection, deals with a seemingly endless variety of bacterial, viral, fungal, and other infectious agents that precipitate an inflammatory response, and has the general task of delineating foreign cells from those that are normal constituents of your body. The brain, on the other hand, remembers your best friend's birthday, calculates the proper tip to leave after dinner, and thinks up excuses to postpone that sigmoidoscopy you have scheduled next week.

Of course we've had hints that the brain and immune system interact at *some* level. Galen, the ancient Greek physician, wrote about the association between personality type (or temperament) and susceptibility to illness. He believed, for instance, that melancholic women were more susceptible to breast cancer than those of sanguine temperament.

In the 1930s, scientist Hans Selye, who was the first to describe the glucocorticoid component of the stress response (see Chapter 3), recognized that manipulations that contribute to this reaction also suppress the immune system. His observations and those of his contemporaries fostered a new appreciation for the way our central nervous and immune systems reciprocally influence each other.

In the 1980s an even more important link was made between the brain and immune system in the discovery of so-called *conditioned immunosuppression.* Behavioral scientists have known since the time of Pavlov that all animals, ranging from your garden-variety snail or fruit fly to a Rhodes scholar, exhibit conditioned learning. The original observations made by Pavlov and his colleagues involved what we now refer to as "classical conditioning".

Let's look at a simple example. Imagine you identify a natural response of an organism to a specific stimulus - say for example, an eye blink in response to a brief puff of air directed at the cornea. This is a fairly straightforward and immediate reaction, right? Things get more interesting, however, if a second stimulus (one that normally does not evoke a response) is presented at the same time the cornea is puffed with air. Consider the example given in Figure 9-1. In this case, a brief auditory tone is sounded at exactly the same moment in time that the air puff is delivered to the cornea. If this pairing occurs several times, an association begins to build up between the two stimuli with repeated exposure. Eventually, the association between the two stimuli becomes so great that a blink response will be elicited by simply sounding the auditory tone by itself. In this context, we say the tone has become a "conditioned stimulus" since it takes on (through conditioned learning) the ability of the unconditioned stimulus (the airpuff) to elicit the blink response.

Classical conditioning goes on at many levels and across all modalities in our everyday lives, most often without our conscious awareness. Neuroscientists are currently identifying the brain circuits involved in this process, although they have proven to be more elusive than was once thought. It turns out that not all forms of classical conditioning are mediated by the same circuitry.

For instance, the eye blink conditioning we have been discussing is dependent on a contribution of processing in the cerebellum, a stalk-like region at the rear base of the brain. A similar type of associative learning known as fear conditioning, in which a noxious stimulus such as a shock is paired with a neutral stimulus (such as a tone or light), does not require cerebellar processing, but is rather more dependent on contributions from limbic structures involved in emotional processing such as the amygdala and hippocampus.

Learning in the immune system

The phrase "conditioned immunosuppression" refers to the ability of a once neutral stimulus to gain influence over the immune system after learning. For example, if an animal is repeatedly given a drug that suppresses its immune system (such as a steroid), and at the same time as the drug delivery is also presented with a small toy, after learning, the toy will begin to elicit a conditioned suppression of the immune system *without the drug*, much like the tone's ability to evoke an eye blink. This is a pretty amazing discovery and has important implications for both neuroscience and immunology.

It shows that it is misleading to characterize the immune response without considering the role of the brain in facilitating the conditioned learning. Before conditioning, the toy is just another harmless and entirely forgettable object that has nothing to do with the immune response. Pair the presence of the toy with immune system suppression, however, and all of a sudden you have a stimulus with direct influence over the function of the immune response. The brain circuitry involved in classical conditioning mediates this transition.

Incredibly, additional experiments have demonstrated that animals with overactive immune systems, which must be treated with immunosuppressive drugs to prevent the development of autoimmune disease, can be successfully treated by a conditioned stimulus in place of the actual drug. In this case researchers substituted the conditioned stimulus (after learning) for the immunosuppressive drug usually given to treat the disorder, and the results were impressive. Both experimental groups – those continually getting the actual drug and those presented with the conditioned stimulus after a brief learning period – showed comparable increases in life span when compared to an untreated group. Thus it has become increasingly clear that, under many conditions, it is almost impossible to truly understand one system without reference to the other.

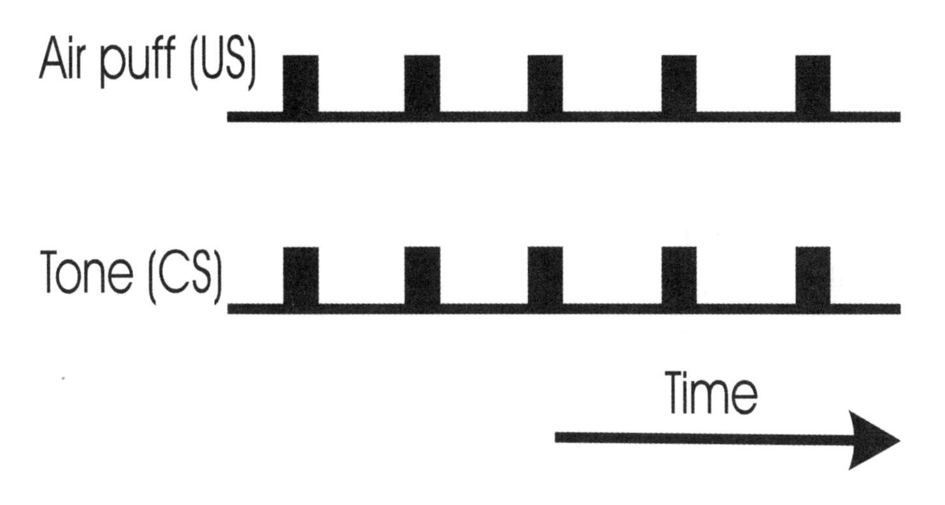

Figure 9-1 – Classical Conditioning – In this example, an unconditioned stimulus (US - a brief puff of air directed at the cornea) is capable of automatically eliciting a blink response. If the air puff is presented several times with a neutral stimulus, in this case an auditory tone, eventually the two stimuli become associated such that the tone presented alone will begin to evoke a blink response. Once this occurs, the tone is said to have become a conditioned stimulus (CS). This is a simple example of associative memory.

In the last two decades the pace of research examining brain-immune system interactions has quickened, with more and more investigators demonstrating intriguing behavioral interactions similar to those discussed above. This pace has gained even greater momentum by recent discoveries showing that the brain has receptors for immune mediators (the chemicals that immune cells use to communicate with each other in the periphery), and that it is also capable of synthesizing these substances.

These observations have yielded tremendous excitement and the possibility for a deeper understanding of how brain-immune interactions contribute to normal and pathological behavior. In the sections that follow, we will examine the key role played by immune function in the etiology of mood and its disorders. We will then go on to explore how the brain-immune system interactions are modulated by stress and the HPA axis. Finally, we will consider the potential avenues for therapeutic intervention of mood disorders via the immune and/or endocrine systems. Before we get too far ahead of ourselves, however, let's start with a peek at how the immune system normally functions.

An Immune System Primer

The main job of your immune system is to protect your body against infection from microorganisms that can come in the form of a virus, bacteria, fungus, or parasite. The tissues and organs of the immune system are scattered hither and thither throughout the body, so communication within this network is incredibly complicated and only partially understood.

The system is composed of lymphoid tissue, fluid known as lymph, and several types of white blood cells that target invading microorganisms. Cells of the immune system congregate in a number of lymphoid tissues including the adenoids and tonsils in the head region, the thymus gland in the chest cavity, bone marrow in the center of long bones, the spleen just beneath the heart, the lymph nodes located under the arms and groin, and several patches in the small intestine (see Figure 9-2).

Immune cells are able to carry out the difficult task of identifying microorganisms and foreign proteins that do not belong to your body, targeting them for removal or destruction, and even forming a memory of the incident.

A foreign substance, which causes the immune system to react, is called an *antigen*. Think of antigens as signal flags (typically composed of proteins, glycoproteins, or carbohydrates) that are carried on the cell membrane of an invading microorganism. They are the clues that allow immune cells to determine if the organism is or is not part of the body – is it one of us or one of them?

There are two broad categories of cell types in the immune system, *phagocytes* and *lymphocytes*, each of which performs a slightly different job. Phagocytes (such as macrophages and neutrophils) are white blood cells that engulf (phagocitize) invading microorganisms, providing the first line of defense against attack. Some phagocytes just hang out in one location like the spleen or lymph nodes waiting to ambush a foreign organism, while others circulate throughout the body looking for invaders. When an infection occurs, chemical messengers released from damaged blood platelets attract circulating macrophages and neutrophils (over here guys) toward the invading microorganism where they ingest the agent, and form the wonderful substance we know as pus.

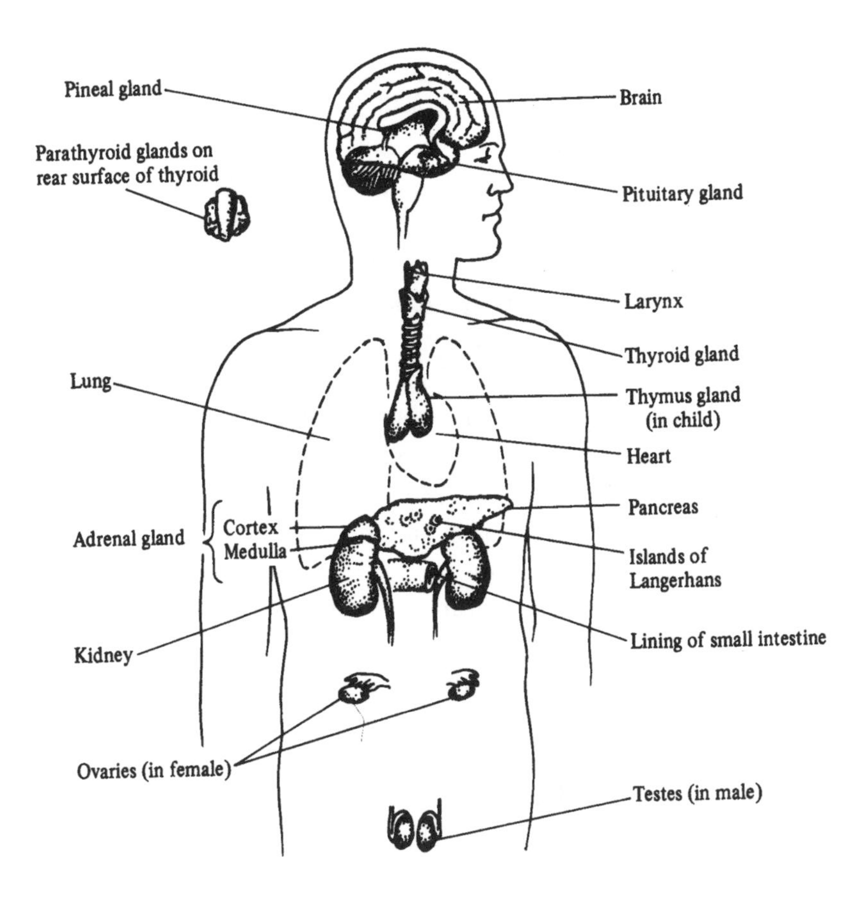

Figure 9-2 – A diagram illustrating the relative locations of the major immune and endocrine organs.

If an infectious agent gets past the process of phagocytosis, a second class of white blood cells called *lymphocytes* provides the next line of defense. There are two classes of lymphocytes: B-cells and T-cells. Both cell types originate in bone marrow, but T-cells migrate to the thymus where they develop into mature immune cells while B-cells mature in the bone marrow (now you know where the "T" and "B" come from).

T-cells and B-cells work together with phagocytes to produce the immune response, but they contribute in very different ways. B-cells carry specific proteins on their cell membrane that are a unique match to a particular antigen. When an infectious agent is encountered with a matching antigen on its membrane, the B-cell is activated and the antigen attaches to its recognition site. The B-cell with the attached antigen then undergoes mitosis and differentiation, which results in the production of two new kinds of cells – plasma cells and memory cells.

The plasma cells deal with the infection directly. They produce an antibody protein that perfectly matches the molecular structure of the antigen in a complementary fashion. The antibodies are then rapidly produced and attach to the antigens on the infectious agent en masse immobilizing it. Once immobilized, our old friends the phagocytes are called into action to ingest the entire yummy antibody-antigen complex. This process is known as the antibody component of the immune response (Figure 9-3).

A secondary immune response occurs via the memory cells. They retain a blueprint for constructing the antibody so if the same antigen is encountered in the future, the response is swift and large in magnitude. We take advantage of this process in immunizing against disease by vaccinating an individual with a weak or inactive version of an infectious agent. This gives the immune system time to construct memory cells that store information pertaining to the antibody. While plasma cells have a very short life span, memory cells can live for years, and if the antigen is encountered again, the memory cells kick into high gear, quickly producing loads of plasma cells with the matching antibody that immobilize the invader and let the phagocytes do their thing.

Antibodies can do only so much, however. For instance, they can inactivate a virus if they get to it before it enters a cell body. But once the virus has infected a cell, its antigen is beyond the reach of antibodies and thus the antibody immune response. So another component of the immune response must take over.

Once a cell has been infected by a virus, a viral antigen is expressed on its surface membrane. Several types of T-cells then become involved in the response, communicating with each other through the circulatory system using blood-borne chemical messengers known as cytokines (such as interleukins, interferons, and tumor necrosis factor). Cytokines are the immune system's equivalent of a neurotransmitter - a chemical messenger that allows one cell to signal or trigger a reaction in a subsequent cell. Once the T-cells recognize the presence of a viral antigen, they proliferate and activate a second class of T-cells called cytotoxic killer cells. Cytotoxic killer cells then destroy the infectious agent without the use of antibodies therefore this process is referred to as a cell-mediated immune response (Figure 9-4).

It should be remarked that this is a vast oversimplification of the different components of immune function. There are normally complex interactions between T-cells and B-cells mediated by the release of cytokines and other immunotransmitters such as interferons and assorted growth factors. For instance, infection by the human immunodeficiency virus (HIV) causes the immune system to collapse by attacking the T-4 helper cell (a type of regulatory T-cell) that normally stimulates the production of B-cells by releasing B-cell growth factor. Thus HIV ultimately interferes with the ability of the body to produce an antibody-mediated immune response, leaving the patient open to infection by any number of invading microorganisms.

Now that we have a basic understanding of the immune system let's examine how its response is modulated by stress and the HPA axis. As we shall see the study of these interactions has led to new theories of mood disorders that may shed light on why these conditions have risen to such high prevalence in the last decade.

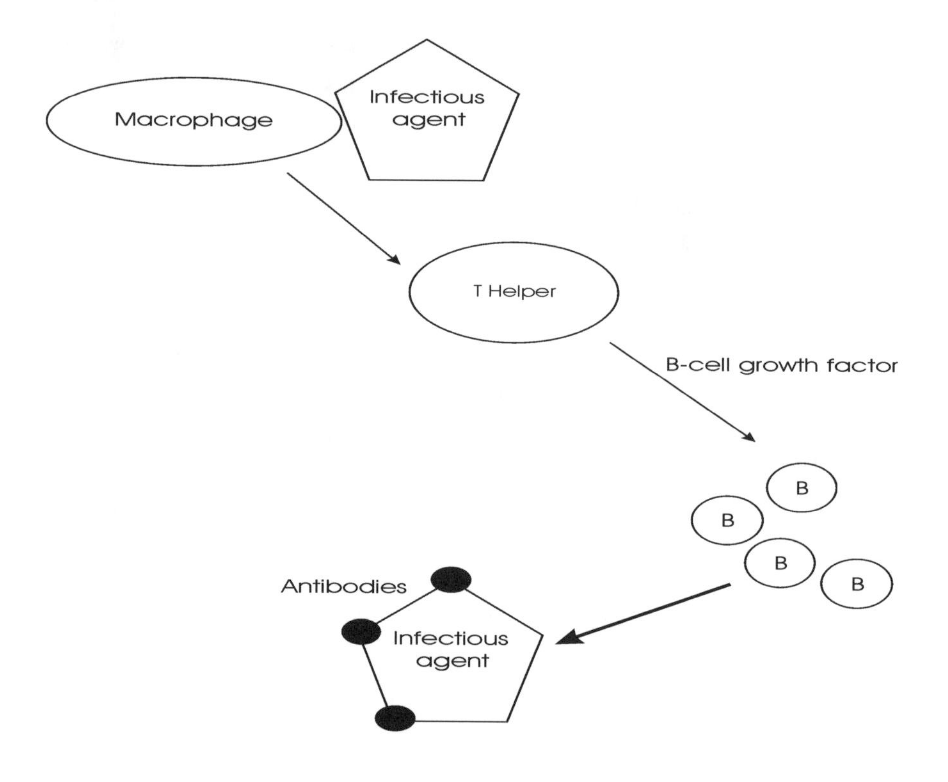

Figure 9-3 – A simplified illustration of the antibody-mediated immune response. A macrophage first detects an infectious agent and releases a proinflammatory cytokine called interleukin-1, which activates T-helper cells. T-helper cells then release a second immune messenger known as B-cell growth factor that stimulates the production of (you guessed it) B cells. B-cells carry specific proteins on their cell membrane that are a unique match to a particular antigen expressed on the surface of the infectious agent. When an infectious agent is encountered with a matching antigen on its membrane, the B-cell is activated and the antigen attaches to its recognition site. The B-cell with the attached antigen then undergoes mitosis and differentiation, which results in the production of two new kinds of cells – plasma cells and memory cells.

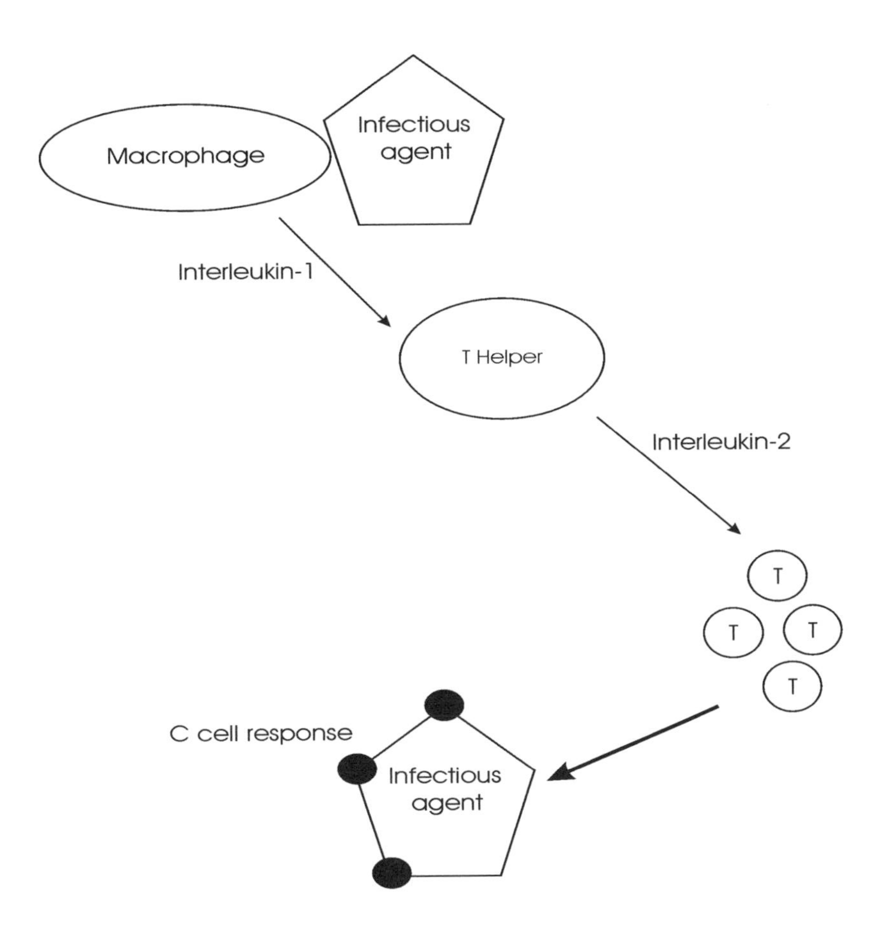

Figure 9-4 – A simplified illustration of the cell-mediated immune response. Similar to the antibody-mediated immune response, a macrophage first detects the presence of an infectious agent and releases interleukin-1, which stimulates the production of T-helper cells. In this case, however, the T-helper cells secrete another cytokine, interleukin-2, which stimulates the proliferation of T-cells. Once the T-cells recognize the presence of a viral antigen they produce a second class of T-cells called cytotoxic killer cells (C) that destroy the invading agent.

Stress and Immune Function

The immune and endocrine systems communicate at multiple levels both directly and indirectly through their mutual connections with the central nervous system, and it has been known for some time that stress can compromise immune function. One of the most replicated findings in this field is the immunosuppressive effect of glucocorticoids.

Glucocorticoids reduce the production of new lymphocytes and impair the release of immunotransmitters like interleukins and interferons so that communication throughout the system is hampered. Glucocorticoids can also enter existing lymphocytes and cause them to synthesize an enzyme that sends them into a programmed suicide mode – resulting in their eventual death. The net result of these effects is a weakened immune system when stress hormones such as glucocorticoids are at high levels of circulation.

Similar observations have been made using externally applied stressors such as electric shock, maternal separation, and social defeat, all of which result in compromised immune function. Likewise, chronic stressors such as living in overcrowded conditions have been shown to suppress both the cell- and antibody-mediated components of the immune response.

As straightforward as this relationship sounds, however, it is now clear that the interaction between stress and immune function is not as simple as we once thought. Clinical studies in humans and experiments using laboratory animals have resulted in a bewilderingly complex set of immune responses to stress that depend on variables like the context in which the stressor was encountered, the specific quantity measured as an index of the immune response (we'll get back to this), the region of the body from which these markers were drawn, and the amount of time passed between encountering the stressor and the assays being taken.

It should be emphasized that the immune response involves a complicated set of reactions that extend over the course of several days. Many immune markers used in these studies as well as those we will discuss in relation to mood disorders are nonspecific and only an intermediate step in the response (e.g. the synthesis of interleukins or mitogen-induced proliferation of lymphocytes), rather than a direct measurement of the immune system's ability to combat infection.

When more sensitive assays of immune function are used, it becomes apparent that many variables change abruptly over time. As illustrated in Figure 9-5, immune function can actually *increase* (based on measures such as calculations of cytotoxic killer cell activity or the concentrations of certain cytokines) after the initial onset of stress, but then decrease back to baseline levels with moderately prolonged exposure. If the stress response system becomes chronically active for an exceedingly prolonged period of time immune system function may then dip below baseline levels (Figure 9-5).

Thus one reason for the difficulty in establishing a simple relationship between stress and immune function is that the latter changes over time in response to the former. This pattern of dynamic change in immune function seems to occur not only in reaction to the stress associated with infection, but also in response to *physical and psychological stressors* that activate the HPA axis.

This is consistent with demonstrations that the release of some interleukins (e.g. interleukin-1) in response to immune system challenge stimulates the hypothalamus to release CRF (remember this stuff? – corticotropin releasing factor), which in turn can induce the release of ACTH and glucocorticoids. It is thought that eventually, with prolonged stress, the increase in glucocorticoid concentrations circulating throughout the blood stream begins to suppress various components of immune function. Thus as you can see, the relationship between stress and immune function is complex.

Matters get even worse when one recognizes that not all interleukins, interferons, and growth factors that are usually considered an index of immune function, respond in the same way to stress. Mood disorders such as clinical depression are often associated with a motley crew of immune system abnormalities, which prevents a simple interpretation. Indeed there is compelling evidence that certain aspects of immune function are weakened during major depressive episodes, while others may actually be facilitated. While reading the next section, keep in mind that we are just beginning to understand the relationship between immune function, the stress response, and psychiatric disorders such as major depression. The data are far from complete, but they suggest exciting new possibilities for deepening our understanding of mood and its disturbances.

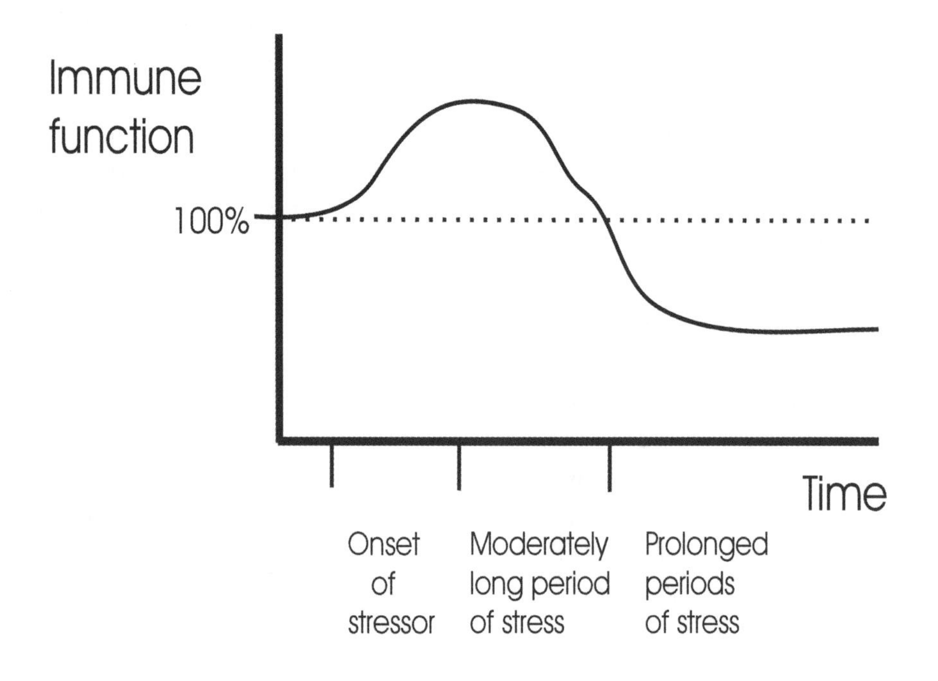

Figure 9-5 – Stress has different effects on the immune system depending on how long it has been present. Immune function, as measured by markers such as cytotoxic killer cell activity or the concentrations of certain cytokines, can be activated in response to brief exposure to a stressor. However, if stress is encountered for a prolonged period of time, an anti-inflammatory response occurs that suppresses the initial activation of the immune system back toward baseline (100%) levels. With chronic exposure to stress, this suppressive trend can continue to reach values well below normal baseline.

Immune System Abnormalities and Mood

There is fairly broad agreement in scientific circles these days that psychiatric illness can compromise immune system function. Most of the studies carried out in the early 1990s showed marked changes in immune response in patients suffering from major depression that include:

1. disruptions in neutrophil phagocytosis
2. reduced lymphocyte proliferation in response to infectious challenge
3. decreases in natural killer cell activity (the cytotoxic response we discussed earlier).

Other studies have considered changes in immune function in people who have in the past or are currently undergoing a great deal of stress such as is precipitated by bereavement, divorce, social isolation, or academic examinations. The pattern of immune system abnormalities that arises in these individuals resembles that which is typically observed in the depressed.

At the same time, however, anecdotal evidence was beginning to pour in from the clinical world suggesting that proinflammatory immunotransmitters such as cytokines may actually *cause* depressed mood. Patients are commonly prescribed cytokines like interferons to treat a host of diseases from cancer to hepatitis and interleukins to treat certain neoplastic diseases. A significant percentage of these folks suffer from side effects collectively referred to as "sickness behavior". The symptoms of this condition include depressed mood, lack of interest in daily activities, suppression of food intake and appetite, sleep disorder, fatigue, confusion, and anhedonia (lack of pleasure in normally pleasurable activities). Do these sound familiar? Moreover, the majority of patients receiving cytokines for immunotherapy who exhibited one or more symptoms of sickness behavior also showed an elevated HPA axis response to psychosocial stressors.

To most scientists and clinicians working in the field, these two sets of data seemed rather inconsistent. On the one hand, we have several markers of immune system function such as decreased neutrophil phagocytosis and natural killer cell activity that suggest a suppression of the immune response in depressives. On the other hand we have a growing number of patients being treated with immune system facilitators like interleukin-1 who report major depression at the onset of treatment that is relieved only after treatment is terminated or switched to an alternative that does not function by activating the immune system.

Although the clinical data were interesting, most scientists had trouble accepting what it implied largely because there was no compelling evidence until the mid 1990s that mood disorders might be associated with *both immune suppression and activation*. Such a theory was first put forth in 1991 by immunologist Robert Smith of the University of Rochester. Based on the clinical data mentioned above and observations that some cytokines cause changes in HPA axis function, he proposed the "macrophage theory of depression", which is illustrated in Figure 9-6.

Smith hypothesized that the abnormal secretion of certain cytokines, namely interleukin-1, interleukin-6, and interferon-alpha, lead to a heightened release of CRF, ACTH, and cortisol that, in turn, contribute to the development of major depressive disorder. In this theory, the immunological abnormalities are the primary or precipitating factor leading to secondary changes in HPA axis function and mood disturbances.

It was not until four years later that the first data demonstrating immune activation in depressed patients was published by psychiatrist Michael Maes and colleagues working at Vanderbilt University. They showed that major depression is not simply associated with immune suppression, but rather with an imbalance that involves activation of the macrophage arm of the immune response along with a relative deficit in lymphocyte functions. This evidence was based primarily on measurements of acute phase proteins and cytokines in the blood plasma of patients suffering from either major depression or treatment resistant depression.

Acute phase proteins are byproducts of cytokine production (particularly of interleukin-1 and interleukin-6) in the early stages of inflammation. They are physiological signatures that indicate the immune system is getting ready for battle. A significant percentage of depressed patients exhibit elevated levels of multiple acute phase proteins such as haptoglobin, neopterin, and alpha-macroglobulin in their blood serum (please don't even try to remember these names).

Consistent with this observation, depressed patients also show elevated blood serum levels of macrophages (immune cells that secrete cytokines), interleukin-1, interleukin-6, and interferon-alpha. So in other words, many depressed patients exhibit several classic signs of immune system *activation*. Proinflammatory cytokine production increases, as do acute phase protein byproducts of this process, along with the number of cells that actually secrete interleukins.

Additional studies conducted in the laboratory of pharmacologist Brian Leonard at the National University of Ireland, an early proponent of the macrophage theory of depression, reported increased phagocytic activity of macrophages in depressed patients. Quite interesting is the finding that the enhanced activation was reversed upon successful treatment with antidepressant therapy.

Together these findings support the notion that facilitated macrophage activity may play a pivotal role in the acute phase of mood disorders or at least modulate the clinical state of depressed patients.

Cytokines and Monoamine Transmission

Some classes of cytokines and acute phase proteins affect monoamine transmission, which may also play a role in generating depressive symptoms. For instance, interleukin-1 directly activates the serotonin transporter. As we have discussed in earlier chapters this transporter protein removes serotonin from the synaptic cleft between cells by reuptake back into the presynaptic cell. The net result is a decrease in available serotonin at the synapse. Selective serotonin reuptake inhibitors like Prozac take advantage of this process by blocking the activation of the transporter thus elevating levels of the transmitter in the synaptic cleft.

Consequently, depressed patients who exhibit an increase in the macrophage arm of the immune response, which includes an increased production of interleukin-1, may also exhibit a reduction in serotonin transmission by an overactive removal by reuptake mechanisms. Indeed, receptors for interleukin-1 exist on serotonin containing brain cells and that both neurons and glial brain cells synthesize these cytokines.

Complementing this finding, proinflammatory cytokines such as interleukin-1 stimulate the activity of a brain enzyme called cyclooxygenase, which in turn stimulates the production of prostaglandins. Prostaglandins are another class of immune mediators (like cytokines) that are significantly elevated in depressed patients. In particular, prostaglandin-2 concentrations are markedly increased in many patients suffering from mood disorder. Drugs that inhibit the production of excessive prostaglandin-2 (such as the cyclooxygenase inhibitor, indomethacin) reduce depressive symptoms in humans and animals that are associated with cytokine-induced sickness behavior.

This is very interesting because prostaglandin-2 is known to be a potent inhibitor of monoamine transmitter release.

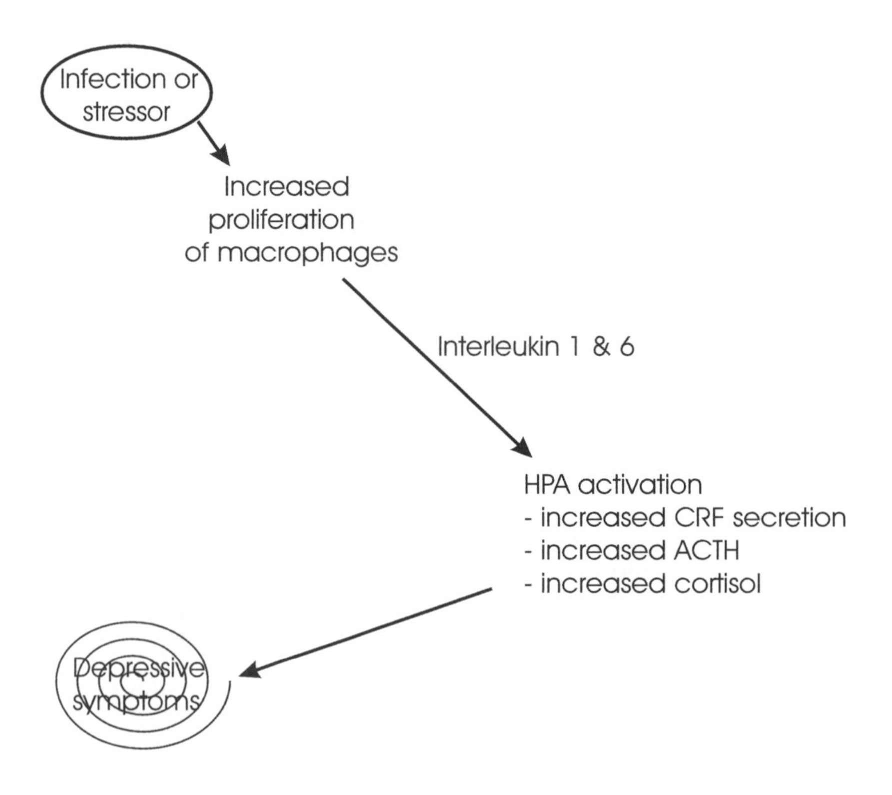

Figure 9-6 – The macrophage theory of depression postulates that an infection or stressor causes an activation of the macrophage arm of the immune response. This consists of increased proliferation of macrophages and the production of proinflammatory cytokines such as interleukin-1 and interleukin-6. Some proinflammatory cytokines stimulate the release of CRF in the hypothalamus leading to a cascade that activates the HPA axis (Chapters 3 through 5). Hyperactivation of the HPA axis and increased levels of circulating glucocorticoids (see text for details) are common markers in depressed patients. This theory offers a foundation through which to understand how immune, endocrine, and central nervous system changes may interact in mood disorders.

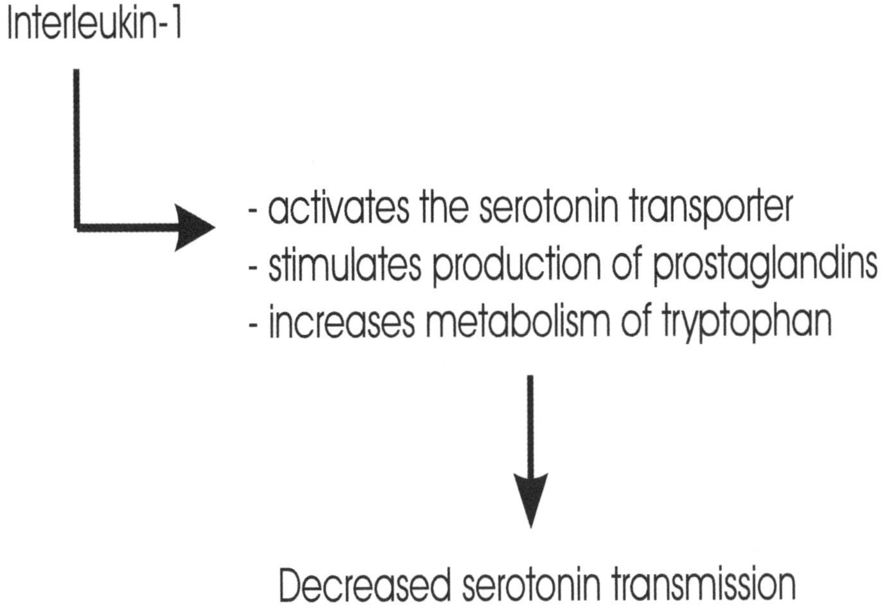

Interleukin-1

- activates the serotonin transporter
- stimulates production of prostaglandins
- increases metabolism of tryptophan

Decreased serotonin transmission

Figure 9-7 – Increased concentrations of proinflammatory cytokines in the circulatory system can potentially modulate monoamine neurotransmission in several ways. For example, interleukin-1 has been shown to activate the serotonin transporter protein. This results in more serotonin being taken back up into the releasing cell. It is also known that interleukin-1 indirectly stimulates the production of prostaglandin-2, an immune mediator that inhibits the release of monoamine transmitters from cells and is elevated in depressed patients. Finally, recent evidence has demonstrated that interleukin-1 activates an enzyme responsible for degrading tryptophan, a chemical precursor required for the production of serotonin. All of these mechanisms thus result in a weakening of serotonin synaptic transmission. Treatment with antidepressants reverses these effects.

The mechanism for this is not currently understood, however, it is fairly clear that when prostaglandin-2 binds to receptors on the terminal endings of monoamine-containing cells, it impedes their ability to release transmitter. Thus both effects of interleukin-1 mentioned above, activation of the serotonin transporter and production of prostaglandin-2, result in *reduced* monoamine transmission (Figure 9-7).

Consistent with this interpretation is the observation that depressed patients exhibit a reduction in serum tryptophan (a chemical precursor of serotonin synthesis) levels. This may be one reason that antidepressant treatment with SSRIs reverses the physiological and behavioral manifestations of cytokine-induced sickness behavior in both humans and laboratory animals.

As interesting as these data are, it is important to realize that not all elevated immune markers in depressed patients have the effect of reducing monoamine transmission. Another link between proinflammatory immune reactions and changes in monoamine transmission can be found in the effects of some acute phase proteins.

While most acute phase proteins that are elevated in depressed patients can be interpreted as having a weakening effect on monoamine transmission based on existing data, there are exceptions. In particular, the acute phase protein, alpha-1-glycoprotein (AGP), is significantly elevated in many depressed patients and is also a known *inhibitor* of the serotonin transporter protein. In fact, AGP is a critical binding protein for many antidepressants that work by selective serotonin reuptake inhibition.

The fact that some immune markers that are known to increase in depressed patients can activate this transporter protein, while others such as AGP seem to inhibit it, illustrates the complexity of the responses these mediators have on central nervous system activity. It also suggests that, as mentioned in earlier chapters, the relationship between monoamine levels and mood disturbances is probably not a simple matter of too much or too little. Most regulatory mechanisms work optimally within a restricted range of values and deviations from this range in either direction can lead to pathological states.

An exciting aspect of the macrophage theory of depression is that it seeks to link observable abnormalities in the nervous, endocrine, and immune systems that are associated with mood and its disorders. Rather than have a monoamine theory of depression, a stress response theory of depression, a viral theory of depression, et cetera, et cetera, the ideas advanced in this chapter are focused toward integrating data from these once disparate fields of investigation into a cohesive and more fundamental framework for understanding and treating these disturbances.

Moreover, the theory has generated novel predictions for treatment strategies. For example, Michael Maes and his colleagues have shown that the concentration of omega 3 fatty acids is significantly decreased in the erthyrocytes (a type of non-lymphoid blood cell involved in the immune response) of depressed patients. This finding is consistent with other reports of an imbalance in the omega 3 and omega 6 fatty acid pathways that can result from increased prostaglandin production. Normally, the ratio of omega 3 to omega 6 fatty acids is tightly regulated by dietary conditions, but proinflammatory immune mediators like prostaglandin-2 have a pronounced effect on this delicate balance.

These findings imply that an increase in omega 6 and/or a decrease in omega 3 fatty acids may contribute to the conditions that cause mood disorders. A diet rich in omega 3 fatty acids (fish oil) may therefore have an immunoprotective effect and some antidepressant properties.

Is Depression an Autoimmune Disease?

In light of the finding that several markers suggest activation of immune function such as increased proliferation of macrophages, elevations in cytokine concentrations, and the appearance of an acute phase response, some investigators have begun to speculate that the pathogenesis of mood disturbances may involve an autoimmune component.

In the early 1990s, data began to trickle in from several research groups indicating the presence of antinuclear, antiphospholipid, and antiserotonin antibodies in depressed patients. So what does this mean? Remember in our immune system primer, we mentioned that antibodies are produced in response to infection. Antibodies are generated to match an invading microorganism with a specific antigen marker. In autoimmune diseases, such as multiple sclerosis and juvenile diabetes, the body produces an immune response focused toward proteins that are mistakenly targeted by the system as a foreign invader. Part of this response can be, in some cases, the generation of antibodies toward the falsely identified "foreign invader", which results in its destruction.

An important point is that some autoimmune disorders have been associated with the production of anti-receptor antibodies. That is, antibodies are actually generated to destroy a particular receptor subtype that normally occurs in the body. The late biologist Anna Sluzewska, one of the earliest proponents of the autoimmune theory of depression, and her coworkers reported in 1997 that patients suffering from major depressive disorder exhibit antibodies against serotonin and receptor gangliosides (a part of the serotonin receptor complex). This suggests that something went awry in these patients driving their system to produce an antibody-mediated immune response against an endogenous protein – the serotonin receptor!

In this series of studies, over 70% of the patients who exhibited antibodies against serotonin or gangliosides were treatment resistant. This means the patients did not display an adequate response to pharmacological intervention with typical antidepressants. This finding suggests an autoimmune component may be particularly important in the pathogenesis of treatment resistant forms of mood disorder.

Patients who had antibodies against serotonin or gangliosides also exhibited elevated levels of the acute phase protein AGP, the cytokine interleukin-6, and decreased serum levels of serotonin when compared to healthy, age-matched controls. This led to quite a bit of excitement and further speculation that deficits in the serotonin receptor system caused by the production of these antibodies, may be associated with the concomitant acute phase immune response that is observed.

A note of caution should be raised here, however, in that there is still no direct evidence suggesting the production of these antibodies actually *causes* mood disorders. As was the case with the endocrine disturbances shown to occur in depressed patients (Chapters 3 through 5), it is currently unknown whether the appearance of antibodies in depressives is a cause or a result of the physiological and behavioral alterations that are observed in the disorder.

The link between viral infections and mood disorders

One possible explanation is that the stimulation of antibodies against serotonin and receptor gangliosides is induced by either a viral infection, or one or more factors associated with the overproduction of cytokines. Several researchers have speculated that viral infections (mostly of the herpes-type viruses) may have a precipitating role in the onset of recurring major depressive disorder.

Antibodies to the herpes simplex virus are elevated in patients with unipolar and bipolar affective disorders. Indeed, a recent study found evidence of active multiplication and elevated antibodies against herpes virus in 41% of patients with major depression[1]. Similar experiments performed in the laboratory of Anna Sluzewska have demonstrated elevated antibodies against herpes simplex virus 1 and 2 in depressed patients during an acute depressive episode. Patients with higher levels of antibodies also showed higher concentrations of interleukin-6 and the acute phase protein AGP.

Another possible route of infection that has been considered recently is the Borna-disease virus (BDV). This infection causes severe disturbances in behavior and cognitive processes that are similar to mood disorders. Interest in BDV has increased sharply in the last few years with discoveries that antibodies against the BDV virus have shown up in several patients with recurrent depression. Furthermore, infectious BDV has been isolated from mononuclear cells of depressed patients, and the BDV antigen has appeared in brain samples from postmortem tissues of depressives.

Although additional studies are needed to clarify the association between these two types of viral infection and depressive disorder, the findings do suggest a close association between an autoimmune reaction and the acute phase immune response in depressed individuals.

When these data are considered in relation to the emergence of particular viral outbreaks worldwide in the last several decades such as herpes simplex virus, it is intriguing to think that this may be one of the contributing factors to the increased prevalence of both unipolar and bipolar affective disorders. To my knowledge, such a comparison employing rigorous temporal and regional constraints has yet to be performed. The central question is whether or not a viral outbreak in a particular location at a particular time is followed by increased rates of mood disorders.

Yet even if an autoimmune factor is found to contribute to the pathogenesis of depression one will be hard-pressed to show it is the sole causal factor. Indeed, the thinking is rather that immune system challenge is one of several elements, interacting with others in the endocrine and central nervous systems, that when combined increase the risk for the onset of the disorder.

For example, consider the following. The production of antibodies against serotonin and gangliosides is associated with an increase in proinflammatory cytokines (interleukins 1 and 6) in depressed patients. These cytokines, in turn, stimulate the secretion of CRF (see Chapters 3 through 5 for discussions of the stress response system) and directly modulate central monoamine neurotransmission. CRF induces the secretion of ACTH and glucocorticoids, which can then regulate the production of the same cytokines. To make matters even more cumbersome, monoamine transmitters and glucocorticoids interact at multiple levels and monoamine transmission (particularly serotonin) is possibly altered by the presence of the antibodies. So we indeed have a morass of relationships compounded by feedback effects between each of the variables.

Although these data are indeed difficult to interpret, they advance the notion that mood and its disorders must be studied at a variety of levels and across traditional system lines if we are to ever have a deeper and more fundamental understanding of the etiology at work. What is the final common pathway among these systems that causes the disease? This is not known at present although we will consider some possibilities in the final chapters.

These interactions, although complex, suggest new interpretations to old data. For instance given the findings we have been discussing thus far, one conceivable idea is that classical antidepressant and antiglucocorticoid compounds may have immunoregulatory actions that contribute to their mood stabilizing properties.

Cytokine Blockers as Antidepressants

Before turning our discussion toward the likelihood of developing cytokine-based therapies for mood disorders, it is worth summarizing the main points we have discussed thus far.

> Cytokines given to humans and laboratory animals induce multiple depressive symptoms that include depressed mood, anhedonia, sleep disorders, reduced food intake, decreased interest in daily activities, hyper-activated HPA axis, and weakened negative feedback regulation of glucocorticoid

levels as shown by a lack of response on the Dexamethasone Suppression Test.

- Exposure to either physical or psychological stress can lead to the production of cytokines in the brain and periphery.

- Patients suffering from depression exhibit an activation of the macrophage arm of the immune response.

- Medical conditions with an inflammatory component are associated with an increased prevalence of affective disorders.

- Antidepressants have anti-inflammatory actions and reverse the physiological and behavioral effects of immune system activation.

Despite these intriguing findings, there are still several outstanding questions that must be addressed in order to extend these results into clinical applications. The first concerns whether or not antidepressants interfere with the proinflammatory response of the immune system.

There have been several models used in recent years to investigate the immune effects of antidepressants. A few studies have been carried out in humans showing a normalization of several immune parameters in depressed patients successfully treated with antidepressants. For instance, a 6-week period of treatment with tricyclic compounds has been shown to reverse the elevation of macrophage production observed at the onset of a depressive episode and result in clinical improvement[2].

Other studies have demonstrated a reversal of elevated cytokine secretion (interkeukin-6) following an 8-week period of treatment with fluoxetine[3]. Similarly, a recent study of men with HIV infection and major depression found that chronic treatment with imipramine attenuated the decline in T-4 lymphocytes production that is a common signature of the disease[4]. This, of course, does not imply that imipramine is a cure for HIV, but rather demonstrates that a classic antidepressant can modulate the lymphocyte arm of the immune response. The question that remains is how? How does imipramine do this?

Because of the nature of these studies it is not possible to conclude whether the effects on immune function are the direct result of antidepressant actions on the immune system or result indirectly from the improved mood associated with treatment. Immune cells are known to have receptors for neurotransmitters, thus antidepressant effects on immune function may be mediated directly at this level.

Additional experiments are needed to elucidate the exact mechanisms governing the relationship between antidepressant treatment and changes in immune system function. These could potentially include a comparison of non-pharmacological treatments of depression, such as electroconvulsive or behavioral therapies, with pharmacological treatments and their ultimate effects on immune function.

Another central question that has been asked recently concerns whether cytokine blockers (antagonists) can ameliorate the symptoms of depression. If proinflammatory cytokines play a *causal* role in the etiology of major depressive disorder, then agents that block this response (by either disrupting their secretion or preventing them from binding to receptors) should have antidepressant effects. There has been surprisingly little research (using humans or laboratory animals) designed to test this hypothesis. This is particularly odd since broad-spectrum proinflammatory cytokine antagonists such as interleukin-4 and interleukin-10 have little or no effects of their own other than interfering with the inflammatory response.

Studies must be conducted that explicitly examine the antidepressant effects of cytokine blockers in order to determine their clinical utility. Indeed, as this book goes to press, there has been only a handful of animal studies conducted to address this question, most demonstrating that interleukin-1 antagonists are able to block the development of depressive and anxiogenic symptoms elicited by maternal separation or other forms of acute stress. This is encouraging, and illustrates that the issue must be addressed in humans as soon as possible since the findings obtained will likely have a significant impact on the field for years to come.

Chapter 10

†

The Immune Response, Stress, and Mood: Is there a Final Common Pathway?

The development of the macrophage theory of depression has had a similar path of discovery to that of the HPA axis proposal described in Chapters 3 through 5. The data driving the theory come from two predominant sources. Preclinical experiments have repeatedly demonstrated that infusing proinflammatory cytokines directly into the brain of laboratory animals produces depressive-like symptoms that include reduced appetite, sleep disorders, decreased sexual activity, anhedonia, fear, and anxiety.

Two additional pieces of the puzzle come from observations in humans. The first is that a significant percentage of patients undergoing immune therapy using proinflammatory cytokines begin to have mood disturbances shortly after treatment onset. Moreover, illnesses or special physiological conditions that result in the increased production of proinflammatory cytokines, such as immediately following the birth of a child for a woman, have been linked to increased rates of depressive disorder.

This is consistent with the second observation that a large percentage of depressed patients show elevated levels of proinflammatory immune mediators such as interleukin-1, interleukin-6, macrophages, several acute phase response proteins, and the appearance of antibodies against serotonin and gangliosides[1].

Since it is well known that several proinflammatory cytokines (e.g. interleukin-1) induce the secretion of CRF, one question that immediately arises concerns whether the depressive effects of these substances are simply a result of their facilitation of HPA axis activity (which is known to occur in many depressives).

Elevated levels of proinflammatory cytokines would be expected to increase the secretion of CRF (Figure 9-6). However, as discussed earlier, depressed patients who exhibit an overactive HPA axis response to stress are also typically glucocorticoid resistant. This means that something has reduced the ability of glucocorticoid receptor activation to down-regulate the further release of the substance through negative feedback (Figure 4-5). As you'll recall this is thought to be the primary reason depressives so often have elevated cortisol levels. Such elevated concentrations will eventually attenuate the further production of proinflammatory cytokines. Indeed glucocorticoid (steroid) administration is one of the most common treatments of inflammatory disease processes. However, there are many important questions about the details of such a mechanism. How fast does "eventually" occur? How large a glucocorticoid response is needed for the attenuation of cytokine production? And so on. We currently have only partial answers (at best) to these questions.

A complementary process by which cytokines may contribute to the etiology of mood and anxiety disorders is through their actions on monoamine neurotransmission (shown in Figure 9-7). For instance, the synthesis of serotonin is limited by the availability of its precursor compound tryptophan. Proinflammatory cytokines have several regulatory actions on tryptophan levels, one of which is through stimulating the production of indoleamine 2,3-deoxygenase, the enzyme that normally metabolizes tryptophan. Thus *elevated cytokine concentrations can indirectly increase the metabolic degradation of tryptophan before it can be used to build serotonin*. This could theoretically result in less available serotonin for synaptic transmission.

As detailed earlier, the presence of antibodies against serotonin and receptor gangliosides suggests that the receptor complex may be targeted in an autoimmune manner. At present, the exact consequences of such a process are unknown, however, one possibility is a reduction of serotonin neurotransmission.

Proinflammatory cytokines such as interleukin-1 have also been demonstrated to directly activate the serotonin transporter protein, consequently increasing the reuptake of the transmitter back into the releasing, presynaptic cell. Again, this could conceivably result in a reduction in serotonin transmission at the effected synapses. Finally, cytokines such as interleukin-1 have been found to indirectly stimulate the production of prostaglandins (e.g. prostaglandin-2), which can impede the release of monoamine transmitters (recall that depressed patients also show elevated levels of prostaglandin-2).

Together these data begin to form a suggestive picture where the immune response to infection or stressors (physical or psychological), or for no obvious overt reason (autoimmune), plays a pivotal role in modulating the endocrinological and neurobiological processes that normally contribute to mood disturbances. But is there a final common pathway? Is there a single common mechanism, shared by all these systems, that is *the* critical cause of depressive disorder? Or might it be more reasonable to think of depression as a syndrome with many possible paths of origination? Some folks may develop anxiety and mood disorders via endocrine malfunction, while others make take the route of autoimmune disturbances. The latter interpretation is encouraged by the fact that not all depressives exhibit the same set of immune, brain, and endocrine marker abnormalities. Furthermore, it is common for a significant subset of depressed patients to be unresponsive to classic antidepressant therapies, suggesting that the targeted monoamine circuits may not play as prevalent a role in the etiology of the disorder in these individuals.

The discovery of cytokines and their receptors in the brain has had an enormous impact on neuroscience. Subsequent indications that brain cytokine transmission can be modulated by peripheral and central immune reactions have led to a new thinking about brain-immune interactions. Add to this the discoveries explicating the manner in which stress influences these two systems, and in turn, the manner in which they affect the endocrine response, and it is no surprise that many scientists in these fields feel we are on the cusp of a deeper understanding of the disorders that are manifested by these circuits.

Of course this goes well beyond mere intellectual curiosity, toward developing alternative therapeutics for combating and preventing the disorders. Some five years ago, the National Institutes of Health asked over 100 prominent scientists to evaluate the current state of knowledge in the biology of mental disorders and suggest directions for future development. The conclusion of their report on trends in neuropharmacology reveals the current excitement in the field.

> "A burgeoning area of research that shows promise for drug discovery stems from work done over the last few years on the neural immune axis. Neuro-immune interactions involve the HPA axis and other neural circuits in the brain that represent potential targets for neuropharmacological intervention… The involvement of interleukins and interferons in behavior and other physiological responses via brain action opens a new and important research arena: the development of pharmacological tools to shape the effects of these substances. Such tools appear to be within our grasp". [From the National Institutes of Mental Health Advisory Committee Report, 1995]

Chapter 11

†

The Neuroanatomy of Mood & Melancholy

O sweet spontaneous earth
how often have the doting fingers of
prurient philosophers pinched and poked thee,
has the naughty thumb of science prodded thy beauty
how often have religions taken thee upon their scraggy knees
squeezing and buffeting thee that thou mightest conceive gods
but true to the incomparable couch of death thy
rhythmic lover thou answerest them only with

spring.

- E.E. Cummings

Close to 400 years have passed since Richard Burton finished his magnum opus, *The Anatomy of Melancholy*. Rarely has a book proven itself to be as unequaled in its richness and depth of description of the disorder to which we have turned our attention. Indeed it is testimony to the inherent robustness of melancholia as a human condition that his book is still popular today.

Contemporary ideas of mood and melancholy have fractured and divided, just as general epistemology has since the renaissance. While Burton was interested in every facet of mood and its disorders from the theological to the medical, our approach will be more conservative. In this chapter, we will attempt to integrate recent observations from immunology and endocrinology with the classic monoamine theories of affective disorder that have been the mainstay of the clinical realm for the last several decades. In order to begin piecing together this puzzle, however, we must first consider the specific brain regions that are thought to be involved in mood and depression and what they may tell us about the neuropharmacology of the disorder.

These days the study of mood and its disorders has become interdisciplinary in nature. What was once primarily the domain of mental health professionals is now investigated by a slew of scientists and practitioners trained in diverse areas ranging from molecular biology to philosophy. And as in other fields of science, newly integrated sub-disciplines frequently emerge that often provide novel insights into old questions. Got a problem that can't be solved by either a clinical psychologist or a neurobiologist? Well then, create new training environments that produce "clinical neuropsychologists".

One such example of this integration can be found in the burgeoning young field of cognitive neuroscience, which as the name implies is concerned with understanding the neural mechanisms that underlie cognitive processes. This area has had explosive growth in recent years largely as a result of innovative technologies developed for imaging brain function in humans in a noninvasive manner. It is now possible to image the dynamic changes in certain markers of brain activity such as regional cerebral blood flow (rCBF), oxygen uptake, or glucose utilization, of a human as they are performing cognitive tasks (such memorizing a list of objects, discriminating between visual patterns, or other "higher-order" processes). These measures of local energy demands across different brain regions indirectly reflect brain cell activation (since it takes energy to activate cells). Other techniques (see Table 11-1) can measure the localization and density of a number of neurotransmitter receptors and transporters (we'll get back to this in a minute).

Modality	Functions	Questions Addressed
PET	Measures rCBF and metabolic rate; maps receptor/transporter density, distribution, and occupancy	Which brain areas are activated during a given task or condition? Are there changes in receptor density, distribution or occupancy during a given task or condition?
SPECT	Measures rCBF; maps receptor/transporter density, distribution, and occupancy	Same as PET but has inferior image resolution.
FMRI	Measures rCBF	Which brain areas are active during a given task or condition?
EEG/ERP	Maps surface bioelectric activity of the brain	Same as fMRI but activated brain area must be inferred from surface recordings.
MEG	Maps surface biomagnetic activity of the brain	Same as EEG/ERP.
MRS	Measures quantitative neurochemistry	

Table 11-1 – Types of functional brain imaging presently used in cognitive neuroscience. Regional cerebral blood flow (rCBF); positron emission tomography (PET); functional magnetic resonance imaging (fMRI); electroencephalogram (EEG); magnetoencephalogram (MEG); magnetic resonance spectroscopy (MRS).

The excitement that followed the development of this new hardware is difficult to overstate, although in the earliest days many studies were plagued by interpretive difficulties. What does it really mean that the prefrontal cortex needs more blood oxygen as compared to other brain areas during a working memory task? Is this effect indicative of inhibitory or excitatory cell activity (both of which have common metabolic demands)? Many investigators saw the new technologies as simply the latest gadgets by which to revisit previously asked questions but with a new twist. All of the old questions were suddenly new again. What happens in the brain when *fill in your favorite classical problem from cognitive psychology*?

Many scientists complained that this focused too much attention on the technology rather than on generating new questions. It's somewhat reminiscent of the proverbial young boy with a shiny new hammer who sees before him an entire world in dire need of pounding. Many of us working in the field were quickly reminded that the hardest part of science is in asking the right question rather than finding a means to answer it. This is certainly true for the experimental data we will discuss shortly related to affective disorders.

Gradually the study of cognitive neuroscience has come into its own. The technology has improved, but most importantly, new theories have emerged that attempt to relate the kind of information one obtains from the imaging procedures with relevant data from other levels of description such as the molecular and behavioral. These theories have helped motivate novel questions that may not have been answerable (or even asked) otherwise.

Table 11-1 shows a partial listing of the functional brain imaging tools used in cognitive neuroscience today along with the neural markers they assay and the typical questions that have been addressed. With regard to our focus these imaging techniques have been used to localize brain regions of synaptic activity associated with a number of psychiatric disorders. Recent studies have proven very exciting in that they have also been able to show how these regions change with the progression of treatment effectiveness.

Techniques such as PET and SPECT are also currently being used to identify the role of specific neurotransmitter receptors and transporters involved in the pathology and subsequent treatment of mood disorders. Indeed new advances are being made every day in the development of labeled tracers for specific receptor subtypes such as those for serotonin-1A and 2A receptors. Let us now take a close look at these data.

Functional Brain Imaging of Depressed Patients

With the advent of each new imaging tool there have been subsequent attempts to use it as a potential classifier or metric of psychopathological states. Since the beginning we have been on a quest to identify what *the* "schizophrenic brain" or "depressive brain" looks like. This approach has met with limited success in that *often the variability in the brain patterns across different states of a single individual is comparable to the variability across a whole population of patients.* This creates quite a statistical problem for isolating a specific pattern related to a pathological condition.

Nevertheless, some regions have been identified that commonly show abnormalities in people suffering from mood disorders. There are, of course, relative degrees of consistency from brain region to region across these data sets. In our attempt to piece together a set of regions or circuits that becomes altered during mood disorders we will begin with those areas that are most often agreed upon to play a role in the illness. We'll then consider how the neuroanatomical evidence relates to the pharmacological theories of mood that have driven our understanding over the last few decades and how they may be changed to better reflect contemporary observations.

Prefrontal cortex

There are several areas within the prefrontal cortex that seem to be affected in patients suffering from mood disorders. Although we currently have a rather limited understanding of the functional properties of the prefrontal cortex, it is clear that it plays a prominent role in so-called executive functions and working memory.

Patients with isolated prefrontal cortex damage resulting from stroke, infection, or trauma often exhibit deficits in short-term memory, decision making, organizing goal-directed behaviors, and tend to perseverate. That is if you train them to perform a task in a certain way, and then switch the rules on them in the middle of the experiment, they are very slow to adapt. Normal subjects typically adopt the new rules over the course of a few trials whereas frontal patients often keep on going with what they have been doing thus far even though their behavior may no longer be adaptive.

Similarly, laboratory animals that have had their prefrontal cortex lesioned or temporarily anaesthetized exhibit comparable impairments across a wide range of experimental tasks.

Electrophysiological recordings are consistent with these functional interpretations in that they have shown single cells in this region become active during these behaviors, suggesting they contribute to the processing of information relevant to the successful completion of the tasks.

One prefrontal region referred to as the subgenual cingulate cortex has repeatedly been implicated in studies of mood and its disorders. Initial experiments using PET imaging indicated that rCBF and metabolism are decreased in the subgenual cingulate cortex of depressed and bipolar patients relative to healthy control subjects. Based on data from magnetic resonance imaging (MRI) and postmortem studies it appears the decrease in activity is caused by a reduction in brain tissue volume restricted to this particular region.

As we have read in Chapter 5, volume reductions can result from a number of pathological processes and mechanisms including a decrease in the number of neurons or glial cells (brain cells that contribute to the structural and nutritional support of neurons), reduced size of either cell type, or decreased dendritic branching of neurons to name just a few possibilities.

It is currently thought that the volume reduction in many depressed patients is caused by a decrease in glial cell numbers along with reductions in the size and complexity of primary neurons. Interestingly, the reduction in volume exists very early in the illness of patients with major depression and bipolar disorder but appears to follow illness onset.

Since it is known that changes in tissue volume affect measurements of metabolic activity such as rCBF, the initial imaging data have since been re-analyzed. After correcting the data for this variable, it now appears that although this region is smaller in depressives, it is actually *more active*. The data adjusted for volume reduction show a substantial *increase* in rCBF and metabolism in the subgenual cingulate cortex of depressed patients as illustrated in Figure 11-1. This interpretation is compatible with recent findings that successful antidepressant treatments result in a *decrease* in metabolic activity in this region in depressives.

This finding is even more intriguing when compared with several studies performed in the last five years that have consistently shown increased metabolic activity in the subgenual cingulate cortex of healthy, non-depressed volunteers during experimentally induced sadness. Once the sadness abates, so to do the changes in regional brain metabolism, which are observed with its induction.

These results suggest an association between mood state and metabolic activity in this brain area. This is congruent with several reports indicating that humans with damage to this region often show abnormal autonomic responses to emotionally provocative stimuli, a reduced ability to experience emotion, and a decreased sensitivity to punishment and reward variables that normally influence behavior.

Relevant to the latter point are data showing that cells in the subgenual cingulate cortex are sensitive to the reward-related significance of a stimulus and also fire in anticipation of reward. Since this region has extensive connections with other circuitry known to mediate reward and pleasurable states, some have suggested that the subgenual cingulate cortex abnormalities seen in depressed patients may contribute to the symptoms of anhedonia (inability to experience pleasure) and impaired motivation that often accompany major depressive disorder. Although additional studies are needed to establish this hypothesis, the idea is very exciting since it may provide a neuroanatomical target for the development of pharmacological and behavioral therapies to alleviate these particular symptoms.

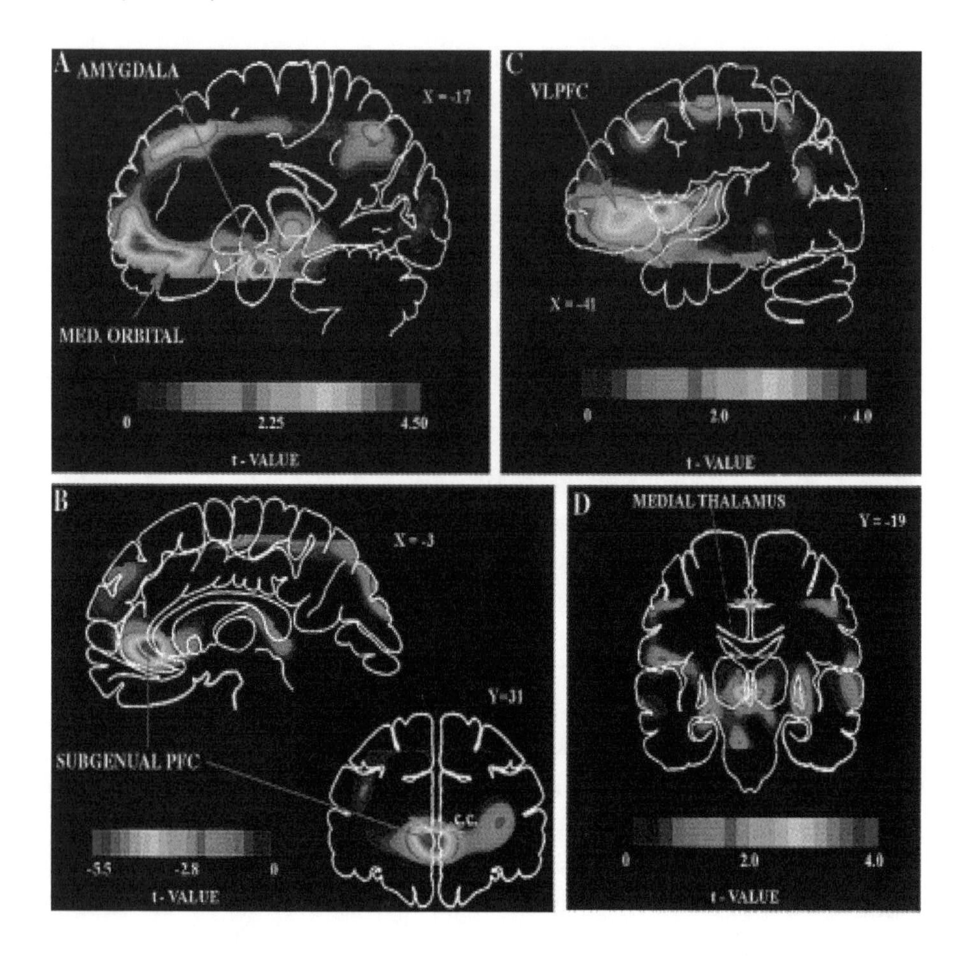

Figure 11-1 – Regions of abnormal regional cerebral blood flow (rCBF) in patients with familial major depressive disorder relative to healthy, aged-matched controls. The relative comparison is made in each panel using a t-value statistic. Higher t-values indicate increased statistical differences in rCBF between the two groups. (A) Bright areas with positive T-values reflect abnormally *increased* rCBF in the amygdala and medial orbital cortex of the frontal lobe. Abnormal activity in these regions has been confirmed in other studies using higher resolution glucose metabolic measures. (B) In this panel, bright areas reflect abnormal decreases in rCBF of depressed patients relative to healthy controls. The largest effect

can be found in a region of the prefrontal area known as the subgenual cingulate cortex. However, this effect is thought to result from a decrease in cortical volume, and when rCBF is adjusted for this variable, local metabolism actually *increases* in this region (See text for details). (C) Other frontal regions including the left ventrolateral prefrontal cortex, the lateral orbital cortex, and interior insula also exhibit abnormal *increases* in rCBF in patients suffering from major depressive disorder relative to healthy control subjects. Adjacent areas including the dorsolateral prefrontal cortex and anterior cingulate show abnormal decreases in rCBF (not shown). (D) Finally, this coronal image shows abnormally *increased* rCBF in the medial thalamus of depressives relative to normal controls. [Reproduced with permission from Drevets, 2000, *Biological Psychiatry*, 48, 813-829.]

The prefrontal cortex of the primate is a complicated place, however, and it should be pointed out that several adjacent areas in this region show an opposing set of results. For instance, several studies have now confirmed *decreased* rCBF and metabolism in the dorsolateral and insular regions of the frontal lobes of depressed patients.

In an exciting series of experiments, psychiatrist Helen Mayberg and her colleagues at the University of Texas Health Science Center in San Antonio, have shown a time-dependent reversal of these abnormalities in depressed patients successfully treated with the antidepressant flouxetine (Prozac)[1]. Brain glucose was measured using positron emission tomography in several patients hospitalized for unipolar depression approximately one and six weeks following the onset of treatment.

The patients showed a pattern of increased metabolism in subgenual cingulate cortex and decreased metabolism in other frontal areas (e.g. dorsal, lateral, and insular regions) at one week, similar to reports obtained in unmedicated depressives. This pattern was completely reversed, however, at six weeks in those patients whose symptoms were alleviated by the treatment. The normalized changes in brain metabolic activity observed at six weeks were uniquely associated with improved clinical state. Those patients that did not respond to drug therapy as measured by the alleviation of symptoms, consistently exhibited the one-week pattern of brain activity.

These data suggest that the adaptive changes in brain patterns observed may underlie successful response to treatment. Indeed two exciting, although preliminary, studies have just been published in the esteemed journal *Archives of General Psychiatry*, which compared pharmacological treatment versus brief interpersonal psychotherapy treatment of major depression[2]. This has been done previously, however, these experiments included SPECT (see Table 11-1 for a description) brain scan images taken before and after treatment of these groups, and compared to non-depressed healthy subjects.

In the study by psychiatrist Arthur Brody and his colleagues at the University of California in Los Angeles, scans were taken before and following either a 12-week treatment course with the SSRI paroxetine or consisting of weekly interpersonal psychotherapy sessions.

Before treatment both depressed groups showed abnormally high metabolic activity in dorsolateral prefrontal cortex, which was normalized back to baseline levels comparable to those observed in healthy subjects following the treatment course.

There are several remarkable points that emerge from this work. First, rectification of abnormal brain metabolism patterns in depressives is seen following successful treatment of major depressive disorder (established using self-reports and the standard Hamilton Depression Rating Scale). Second, both treatments (drugs or psychotherapy) resulted in an alleviation of symptoms and both treatments produced a similar set of changes to the abnormal brain patterns. Thus although the treatment methods were entirely different, they resulted in a comparable set of adjustments in the patterns of regional metabolic activity in the brain.

Does this mean that both treatments are doing the same thing in the brain? Probably not, however, the end result of both processes may be quite similar at the rather macroscopic scale of regional cerebral blood flow. In other words, different cellular effects, mediated by different treatment regiments, may lead to similar changes in gross brain activity. From a therapeutic perspective one hopes this is indeed the case since it suggests that different forms of therapy may be used to achieve the same desired end result - normalization of abnormal regional metabolic activity along a set of specific brain circuits. Isolating these circuits and identifying the functional connectivity that changes during depression and with the therapeutic actions that lead to recovery may be the first step in determining the final common pathways involved in successful treatment.

Additional studies are needed to further understand the role of the prefrontal cortex in mood and its disorders. From the behavioral, electrophysiological, and lesion data discussed thus far, abnormalities in this region may account for several of the symptoms known to be associated with major affective disorder including anhedonia, lack of motivation, working memory impairments, and decreased adaptability/ flexibility.

Limbic structures

The prefrontal cortex has reciprocal connections with a host of other brain regions thought to be involved in regulating mood and/or contributing to many of the symptoms of major depressive disorder. Several of these brain areas such as the amygdala, hypothalamus, hippocampus, raphe nuclei, and locus coeruleus, have already been discussed so we will only return to them briefly here.

The two most prominent limbic structures consistently hypothesized to play a role in the regulation of mood are the amygdala and hippocampus[3], shown in Figure 11-2. Both regions are to some extent important for the processing of emotional stimuli, particularly when it must be encoded into long-term memory.

An elaborate circuit has evolved connecting these two structures with midbrain areas including the locus coeruleus, raphe nuclei and ventral tegmental area, which originate monoamine-containing cells. Several experiments have now demonstrated, that the cellular and molecular mechanisms thought to support long-term memory are modulated by these transmitters. Furthermore, experimental manipulations that alter the availability of monoamine transmitters in the amygdala and/or hippocampus have repeatedly been shown to disrupt the acquisition of long-term memory and other putative functions of these regions.

There is quite a bit of evidence that these two areas are intimately involved in *associating the emotional (or affective) content of a stimulus with its other descriptive sensory features.* For instance, suppose you're walking down a dark street late one evening when you suddenly see a rough looking group of nuns. You've been scared the death of nuns since childhood so immediately your hypothalamus activates the HPA axis and the sympathetic division of your autonomic nervous system, which crank up the fight or flight response – virtually every organ in your body is affected. One open question about this process is how the sensory information entering the brain (the physical appearance of the nuns) becomes linked up or associated with this particular emotional reaction. There is compelling evidence that limbic areas are involved in this process, particularly the amygdala. One function of the region then, may be to associate the emotional features of an object/situation with other descriptive features that are present.

Back in the 1930s neuroscientists Heinrich Kluver and Paul Bucy working at the University of Chicago provided the first clues that this may be the case. They found that monkeys with damage to their amygdala and adjacent regions of the temporal lobe exhibited a collection of strange behaviors including psychic blindness, oral tendencies, altered sexual behavior, and emotional changes. Collectively, these behavioral changes are referred to as Kluver-Bucy syndrome.

Although the monkeys had normal visual functioning, they did not seem to *recognize* familiar objects or understand their meaning (hence the term psychic blindness). Following additional experiments it became clear that many of the recognition deficits were centered on an inability of the animals to interpret the usual emotional component of an object. In other words, if successful completion of the recognition task required the use of the emotional features of an object, performance deteriorated rapidly. For example, monkeys are usually fearful of snakes and avoid them whenever possible. Kluver-Bucy monkeys, however, no longer displayed signs of fear (physiological or behavioral) when encountering snakes, even after being bitten.

Humans with isolated damage to the amygdala as a result of stroke or other medical conditions such as Urbach-Wiethe disease, exhibit symptoms very similar to those Kluver and Bucy observed in their monkeys. Most notably people with this type of damage often display a blunted emotional experience to normally provocative stimuli. Additionally, they often show marked impairments in recognizing the emotional expressions of others. Thus both the animal and human data suggest that the amygdala has an important role in associating an emotional label to a given object or situation. Clearly, this type of function is critically important for the survival of an organism.

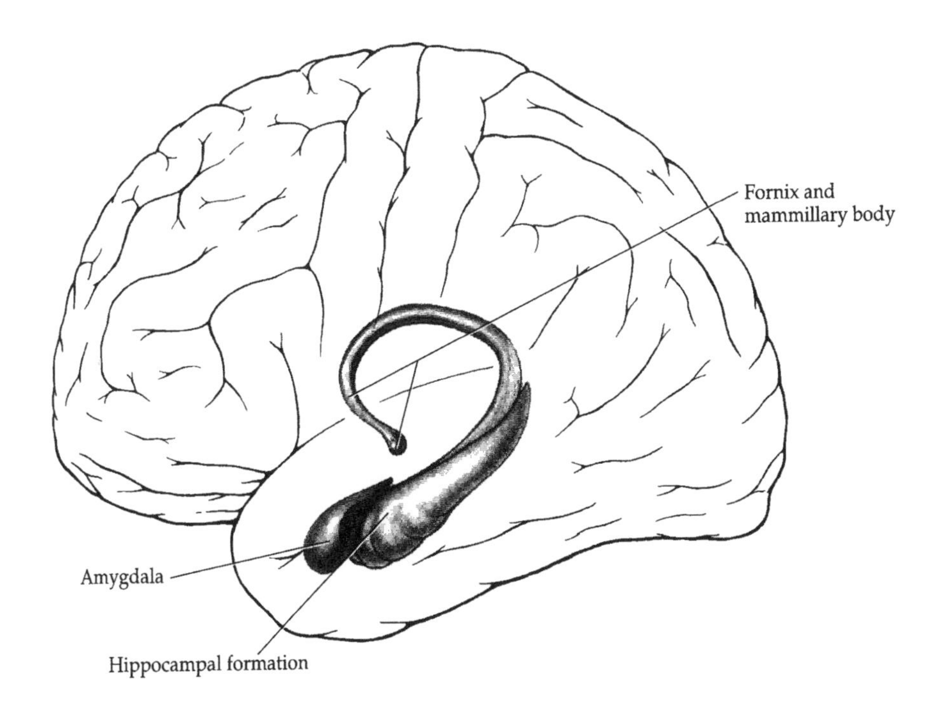

Figure 11-2 – Shown are the relative positions of the amygdala and hippocampus in the medial temporal lobe area of the limbic system. The front of the brain is to the left. [Reproduced with permission from Martin, 1996, *Neuroanatomy Text and Atlas*, Appleton and Lange.]

Contrasting this, electrical stimulation of the amygdala (which has an effect of activating the cells in this region) produces anxiety, fear, dysphoria, the recollection of emotionally provocative events, and an increase in cortisol secretion. Several studies have shown rCBF and glucose metabolism are abnormally high in the amygdala of depressed patients. Incredibly, the magnitude of this increase correlates positively with depression severity, such that individuals with the greatest amygdala activation display the strongest clinical symptoms.

Glucose metabolism and rCBF levels are reduced to normal levels in patients following successful treatment with antidepressants such as

SSRIs. This result is congruent with evidence from laboratory animal experiments showing that antidepressants have an inhibitory effect on cellular activity in the amygdala. As you'll remember from Chapter 4, amygdala stimulation also leads to the increased secretion of CRF from the hypothalamus. Thus, one interpretation of these converging data is that the amygdala has a role in associating the emotional features of an object or situation with the stress, immune, cognitive, and behavioral reactions that are called for.

The behavioral data suggest that excessive amygdala activation may produce the feelings of anxiety and panic often encountered in clinically depressed patients. This increased activation may also over-excite downstream regions the amygdala normally communicates with such as brain stem nuclei involved in the regulation of sleep-wake cycles, social behaviors, and pain sensitivity. Consequently the amygdala has both functional and anatomical properties that allow it to influence a number of the physiological and behavioral processes disrupted during a mood disorder episode.

The hippocampus also contributes to some of these disturbances. Many patients suffering from a major depressive episode exhibit increased rCBF and glucose metabolism during this period that normalizes back to healthy levels following a 6-week course of successful treatment with antidepressants[1].

Similar to the effects seen in the amygdala, this reduction in hippocampus activity was observed only in those patients who reported an improvement in clinical symptoms. This finding is consistent with experimental studies that have shown the hippocampus is inhibited by serotonin. Most often this has been demonstrated by either infusing serotonin directly into the hippocampus and measuring cellular population activity or by stimulating the raphe nuclei, which consists of serotonin-containing cells that release the transmitter into the hippocampus. There are a number of possible biophysical explanations for how serotonin reduces amygdalar and hippocampal activity but we will not go into these at this time. Suffice it to say that it is now becoming more and more accepted that the therapeutic effects of antidepressants are, in part, associated with a reduction in activity at several limbic brain sites.

The recent development of new radioactive tracers specific for serotonin-1a and serotonin-2a receptors that can be used with PET scanners has for the first time made it possible to image serotonin transmission in depressives. Using these tracers, scientists at Yale University and the National Institutes of Mental Health recently demonstrated a selective decrease (downregulation) in serotonin-1a receptor density in the hippocampus of depressed patients relative to healthy control subjects.

Since serotonin and glucocorticoid receptors are often co-localized in the hippocampus, this raises the possibility that hyperactivation of the stress response in these patients and the subsequent excess of cortisol contributes to their downregulation. Because they are assumed to be postsynaptic receptors, their downregulation will result in a reduction of serotonin transmission (that normally inhibits hippocampus activity) and thus an *increase* in hippocampus activity that is observed during a depressive episode.

This pathology may also be related to the so-called glucocorticoid resistance seen in depressives. When glucocorticoid receptors located in the hippocampus are activated, a cascade begins that results in a negative feedback signal being sent to the hypothalamus, anterior pituitary, and the adrenal gland to ease off on the HPA axis response. In other words, the hippocampus detects excess cortisol and says, "enough already, we have way too much of this stuff".

This regulatory signal does not work properly in the majority of depressives, leading to persistent glucocorticoid production. The HPA axis keeps pumping away since no signal is coming back from other HPA sites or limbic regions telling it to slow things down. The fact that SSRIs have a restorative effect on this regulatory signal taken along with the results we have just discussed, serves to deepen suspicions that one of the central points of pathology leading to mood disorder symptoms may reside in a dysfunctional coupling of the serotonin-1a and glucocorticoid receptor in several limbic regions.

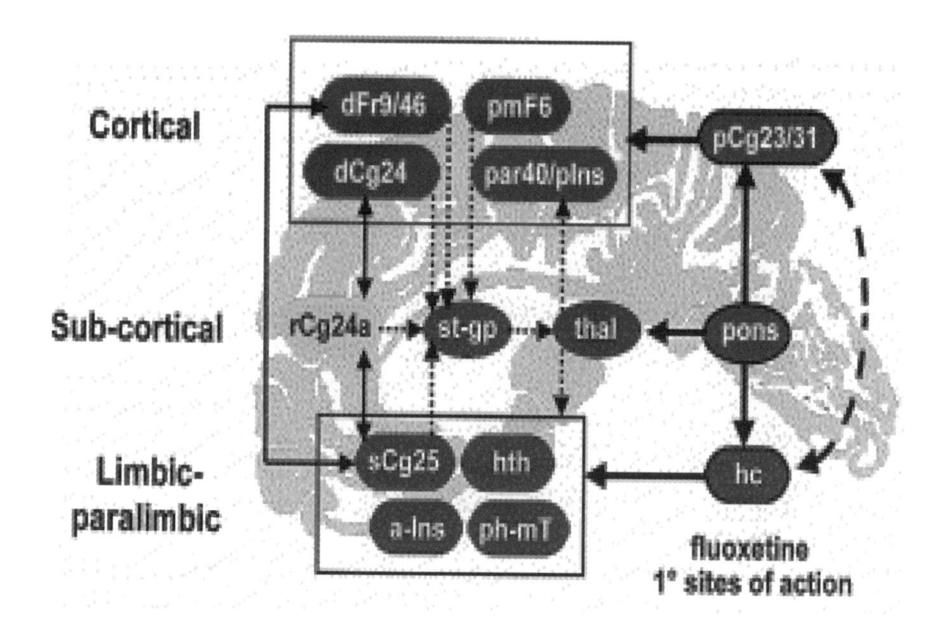

Figure 11-3 – Response-specific effects of a 6-week treatment exposure to the SSRI flouxetine are illustrated in this schematic model. Structures with known anatomical and functional connections that exhibit significant metabolic changes in depressed patients following a responsive (symptom alleviating) 6-week treatment course are grouped into three components – cortical, subcortical, and limbic/paralimbic (compare with Figure 11-1). The solid black arrows indicate known reciprocal cortico-limbic, limbic-paralimbic, and intracingulate anatomical connections. The thin dotted lines signify known cortical-striatal-thalamic pathways. In this model illness recovery is associated with a normalization of the abnormalities in several key brain structures (Figure 11-1). This normalization involves a decrease in metabolic activity in multiple limbic/paralimbic (e.g. hippocampus, hypothalamus, and amygdala – not shown) and subcortical (e.g. thalamus and striatum) sites along with an increase in rCBF in previously hypoactive brainstem (e.g. raphe nuclei of the pons) and some cortical (e.g. posterior and

anterior cingulate cortex, dorsolateral prefrontal cortex, and posterior-parietal groups) locations. It is further hypothesized that the initial loci of the flouxetine-induced changes occur in the hippocampus, raphe nuclei (site of serotonin-containing cells), and perhaps the posterior cingulate cortex. Abbreviations are based on Brodmann's designations. dFr9/46 – dorsolateral prefrontal cortex; dCg24 – dorsal anterior cingulate; par40/pins – inferior parietal/posterior insula; pCg23/31 – posterior cingulate; rCg24a – rostral anterior cingulate; st-gp – striatum and globus pallidus; thal – thalamus; pons – raphe nuclei of the pons; Cg25 – subgenual cingulate cortex; hth – hypothalamus; a-Ins – anterior insula; ph/mT – parahippocampus/ medial temporal lobe; hc – hippocampus. [Reproduced with permission from Mayberg et al., 2000, *Biological Psychiatry*, 48, 830-843.]

Summary of anatomical findings

There are of course other regions that have shown abnormalities in depressed patients, but with much less consistency from study to study. These include motor areas like the cerebellum and striatum, and portions of the medial thalamus, a region involved in relaying sensory information from subcortical structures up to the neocortex for more complex information processing.

Based on the limited anatomical data we have from PET and fMRI imaging studies a model of treatment response has emerged that suggests the subgenual cingulate cortex, amygdala, and hippocampus are the first to be normalized with antidepressant therapy. As illustrated in Figure 11-3, these limbic/paralimbic structures are the primary regulatory regions that show initial changes leading to recovery from major depression and are perhaps the earliest involved in the recurrence of an episode.

As can be seen from the figure, these areas are anatomically poised to have a tight regulatory control over other regions that contribute to the regulation of mood including the hypothalamus and HPA axis, brainstem sites that control the release of monoamine transmitters, and a large number of cortical regions that participate in higher cognitive processes.

In this model illness remission is closely associated with an inhibition of limbic/paralimbic structures like the hippocampus and amygdala and the simultaneous activation of previously hypofunctioning neocortical areas. Because it is anatomically positioned to facilitate communication between limbic/paralimbic and neocortical structures, and is consistently implicated by imaging studies, the subgenual cingulate cortex is thought to play a pivotal role in driving the neurobiological components of mood disorder.

One hypothesis is that pathology begins in this region and spreads quickly through reciprocal connections with other structures. The extent and locations of these affected brain areas suggest possible mechanisms to account for many of the symptoms associated with depressive disorders such as those involving disturbances in mood, motor behavior, and cognitive functioning.

- Abnormalities in motor areas such as the striatum and cerebellum, when associated with those reported for the thalamus, may contribute to deficits in a number of psychomotor functions such as movement planning, motivation, and goal-directed behaviors.

- Increased activation of subgenual cingulate cortex with reciprocal decreases in insular and dorsolateral prefrontal cortex may contribute to working memory deficits commonly linked to mood disorders.

- Abnormally high activation of the amygdala, hippocampus, and related limbic areas, taken in conjunction with emerging theoretical ideas for the roles they play in long-term memory and emotional processing, suggest a mechanism by which these functions are altered in major depression.

- Additionally, several of the limbic and prefrontal structures affected in depressed patients are known to be part of a so-called "reward circuit" involving the neurotransmitter dopamine and multiple subcortical nuclei including the ventral tegmental area and lateral septum (also known to be affected in depression).

In the next chapter we will examine how this circuitry overlaps with the regions involved in mood disorders and consider therapeutic implications. A reasonable question to ask is whether abnormalities in the reward circuit may lead to the most prevalent depressive symptoms: (1) an inability to find pleasure in activities that are usually enjoyable; and (2) a lack of motivation for daily activities. Some imaging studies have found abnormalities in several brain sites that are part of this circuit (e.g. the ventral tegmental area and periaqueductal gray area), however, others have not. This variability across studies may result from any number of processes.

One possibility is that this circuit malfunctions in only a particular subtype of mood disorders, and the effect is smeared by including both subjects who do and do not fall into this category. The problem is that we do not yet know how many subtypes of depression occur (this is the old issue of identifying the correct phenotype – Chapter 1).

Another possibility is that there is indeed a final common pathway that malfunctions to cause major depressive disorder, but many different routes to this destination. In order to address these alternatives we must continue to gather data by employing tightly controlled imaging studies of depressed patients. Most importantly as our basic knowledge of neuroscience increases we will be better poised to make intelligent hypotheses about brain area involvement in the etiology of specific symptoms. In this manner phenotype may be defined by the convergence of supporting data at multiple levels of observation (i.e. molecular, physiological, and behavioral). The hope is that this will shed light on both the basic etiology of mood and its disorders as well as provide a focus for the rational development of drug and behavioral therapies that target the systems involved.

Chapter 12

†

pleasure, Pleasure, PLEASURE...!

In 1953 James Olds and Peter Milner working at the California Institute of Technology in Pasadena made an amazing serendipitous discovery. They showed that similar to natural rewards such as food, water and sex, electrical stimulation of certain brain sites (and only those sites) can be used to condition animals. In their initial experiments Olds and Milner observed that rats seemed to develop a preference for being in portions of a testing chamber where they knew they would receive the brain stimulation.

Conditioned place preference, as this is often called, is taken as an index of drive or motivation to receive a pleasurable reward. For example, consider the following experimental scenario. Imagine we have a testing arena divided into two equal sides, with each being visually distinct such that a laboratory rat can discriminate between the two. Further suppose that our rat is hungry and consistently given food when in Side A and nothing in Side B. It doesn't take a genius to see that when the rat is hungry in the future, if given a choice between the two sides, it will tend to spend more time in Side A where it received food before. The rat develops a *preference* for a specific place in the test chamber.

In this example the primary factor driving the conditioning is the motivation to reduce hunger. Rats and other animals (including humans) are conditioned by a variety of stimuli even when there is no need to relieve an aversive state (such as hunger), but rather in order to obtain something experienced as pleasurable.

If given a choice between Side A, which has 35% sucrose-sweetened water and Side B, which has 10% sucrose-sweetened water, which side do you think our rat will prefer? It turns out that even well hydrated/fed rats prefer the sweeter water. In this example, one might argue that the factor driving the conditioning is the motivation to obtain something pleasurable. Olds and Milner showed that the same kind of conditioning (with similar learning and extinction rates) takes place with direct brain stimulation, presumably of the circuit that is ultimately activated by naturally rewarding stimuli.

In an interesting twist they arranged a new experiment such that the rat could *stimulate itself* by pressing a lever positioned in one of the corners of the testing chamber. Being inquisitive by nature, once a rat is placed into a novel environment it tends to explore the space – peeking here, poking there. Eventually it will accidentally push the lever and subsequently receive electrical stimulation. What do you suppose the rat does next? You guessed it – it keeps pressing that lever like there's no tomorrow. In fact animals like to self-stimulate (their reward circuit) so much that they'll often do so to the exclusion of food, water, or access to sexual partners. Indeed, it is common that a rat will press a lever repeatedly for stimulation to the point that lactic acid build-up in the muscles is the only thing that prevents further responding.

This discovery raised so many difficult questions. What is the rat experiencing during the stimulation? Does the rat find this pleasurable? Does this circuit mediate behaviors motivated by natural rewards? Is the circuit involved in addiction?

These experiments have since been replicated hundreds of times, in many species including humans, who often describe such stimulation as having a sexual quality. Subsequent experiments have delineated a complex circuit of interconnected areas that result in self-stimulation including the hypothalamus, amygdala, hippocampus, anterior cingulate cortex, frontal cortex, ventral tegmental area, and, an area we have not yet discussed called the nucleus accumbens (Figure 12-1).

The nucleus accumbens receives information from a number of structures listed above including the hippocampus, amygdala, ventral tegmental area, and prefrontal cortex, and passes it on after further processing to several areas involved in producing a motivated response such as the striatum, hypothalamus, and multiple midbrain cell groups. Thus the nucleus accumbens is in a pivotal position to integrate information that stimulates limbic and neocortical pleasure centers with an appropriate behavioral response.

An additional element common to many of the brain areas that mediate self-stimulation is that they receive connections from dopamine-containing cells in several midbrain regions including the ventral tegmental area (VTA) and the substantia nigra (the latter is implicated in Parkinson's disease).

When cells fire in the VTA they release dopamine (a monoamine transmitter) into many of the structures that comprise the brain reward circuit. Interestingly, self-stimulation is disrupted in these structures when drugs that block dopamine (e.g. antagonists like haloperidol) are administered. Contrasting this, if given the chance animals will self-infuse drugs that activate dopamine receptors, with a pattern of behavior and locations similar to that seen with electrical self-stimulation. In other words, the common point that seems to drive the sensation of "pleasure" is the binding of dopamine to receptors in these select brain structures. Block this from occurring and self-stimulation disappears.

We now know that many drugs with addictive potential facilitate dopamine transmission. For instance, ingestion of cocaine or amphetamines results in increased dopamine release into many of these target areas, which presumably increases the likelihood of receptor binding. Thus it is no surprise that many scientists studying the biology of addiction have generally focused on this circuit, and dopamine transmission in particular.

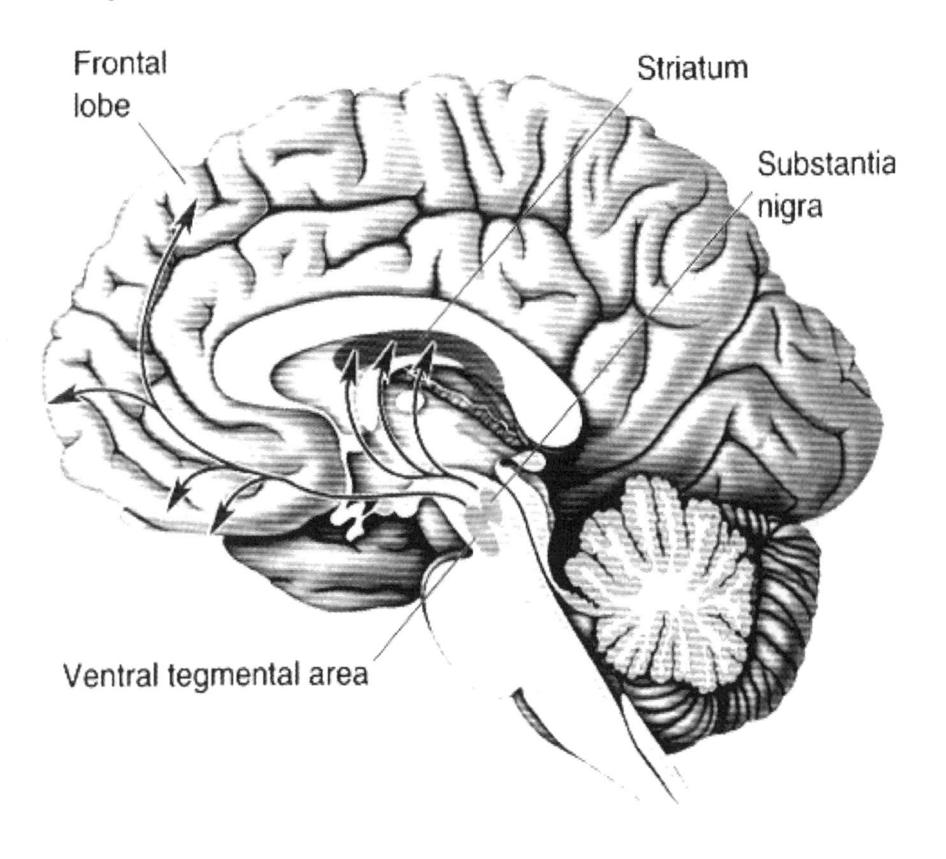

Figure 12-1 – The dopamine system originates predominantly from two cell groups located in the midbrain called the ventral tegmental area (VTA) and the substantia nigra. These cell groups project to target structures in the limbic system, the striatum associated with motor behavior, and various cortical sites in the frontal lobe. [Reproduced with permission from Bear, Connors, and Paradiso, 2001, *Neuroscience: Exploring the Brain*, Lippincott, Williams, and Wilkins Publishers]

Dopamine Transmission and Mood

As the studies of dopamine and the reward circuit continued some investigators began to question whether it might play a general role in regulating mood. There were essentially two lines of evidence that prompted this suggestion.

First, some drugs that increase dopamine levels, such as cocaine and amphetamine, can elevate mood, albeit for a very short period of time. Conversely, drugs that disrupt dopamine transmission by either reducing transmitter levels (e.g. reserpine – Introduction) or blocking receptor binding (e.g. some antipsychotic medications) often produce depression. This suggests a connection between dopamine transmission and mood.

At the same time these issues were being considered there was a push within the mental health community to provide a classification system for psychiatric conditions that resulted in the Diagnostic and Statistical Manual of Mental Disorders (DSM) used throughout North America today. Among the primary symptoms it lists for major depressive disorder are: (1) an inability to find pleasure in activities that are usually enjoyable; and (2) a lack of motivation for daily activities; which certainly sound like they may (at least in part) be mediated by the dopamine reward circuit. Thus in the mid 1970s the first dopamine theories of depression were developed. The logic being that perhaps this circuit malfunctions in depressives resulting in a disruption of behaviors motivated by reward that depend on the ability to experience pleasure. This hypothesis implies that antidepressant effects may be mediated by improving dopamine transmission within this circuit.

The first studies that explicitly examined the effects of antidepressants on dopamine transmission were conducted in the late 1970s, primarily in the laboratories of the Italian pharmacologists Gino Serra at the University of Cagliari and Gian Luigi Gessa at the University of Sassari. They took advantage of the fact that dopamine receptor agonists (activators) like apomorphine, when given in small doses, reduce locomotor activity and produce sedation (at high doses these drugs have just the opposite effects – hyperactivity and insomnia). They discovered that chronic antidepressant treatment potentiated the hyperactivating effect at high doses and eliminated the sedating effect seen at low doses.

These results suggest that chronic antidepressant treatment facilitates dopamine transmission.

This effect has been replicated many times (but not in every study performed) in the last twenty years using a number of different antidepressants and non-drug therapies for treating depression such as electroconvulsive shock and REM sleep deprivation. Most of the evidence to date suggests that chronic antidepressant treatment facilitates dopamine transmission, resulting from an increased sensitivity of postsynaptic dopamine receptors (upregulation). This may be due to either the production of more receptors or an increased sensitivity of existing receptors – it is still unclear.

From subsequent behavioral and *in vivo* microdialysis (a technique used to measure the concentration of a specific neurotransmitter) studies it appears that the potentiation of dopamine transmission is preferentially observed at limbic, striatal, and prefrontal cortex locations. Indeed a few studies have even reported that chronic administration of SSRIs like flouxetine results in increased dopamine release into several limbic and prefrontal cortex structures. Thus it may be the case that *both transmitter release and transmitter binding* are facilitated by antidepressants.

These results caused quite a bit of excitement for those studying the biology of mood. Additional experiments soon demonstrated that reward-related behaviors that are normally facilitated by dopamine agonists were also facilitated by chronic antidepressant treatments. These findings provided an additional link between dopamine transmission in the reward circuit and the regulation of symptoms such as anhedonia and the lack of motivation commonly observed in depressed patients.

Consistent with this interpretation recent experiments have shown that some drugs that enhance dopamine transmission may have direct antidepressant properties. Both chronic and acute treatment with selective dopamine-1 receptor (a dopamine receptor subtype) agonists has an antidepressant effect similar to classic therapies like imipramine in animal models of depression (e.g. brief stress, maternal separation, and learned helplessness models).

These compounds have been shown to reverse the depressive state induced in these models just as well as classic antidepressants. This implies that enhanced neurotransmission at dopamine-1 receptors may have antidepressant actions. It is very exciting to note that this hypothesis is just beginning to be tested in human patients suffering from major depressive disorder. In fact, the dopamine agonist nomifensine (known in the U.S. as Merital), which also has an ability to block the norepinephrine and serotonin re-uptake transporters, initially showed great promise as an antidepressant, but was quickly taken off the market when it was discovered that it can, in rare instances, precipitate the blood disorder haemolytic anaemia. However, new compounds with greater specificity for activating only dopamine receptors (and hopefully minimizing unwanted side effects) such as the selective agonists pergolide and pramipexole, have been tested with encouraging results. At present, additional clinical trials are needed before the U.S. Food and Drug Administration (FDA) approves the licensed use of these compounds.

Chapter 13

†

Biological Rhythms and Mood

It's often said that nature loves rhythms. Creatures large and small, ranging in biological complexity from single-celled amoeba to primates, display rhythms on a variety of spatial and temporal scales.

For instance, almost all land-dwelling animals exhibit a host of physiological and behavioral phenomena that are regulated across an approximately 24-hour period. We call these *circadian rhythms* from the Latin terms *circa*, "approximately" and *dies*, "day". Examples of circadian rhythms include daily fluctuations in core body temperature, blood flow, urine production, the secretion of hormones such as cortisol and growth factor, metabolic rate, activity level, alertness, and sleep-wake cycles to name just a few.

Although nature is good at keeping time, it is not perfect. Each day these and other rhythms have to be reset by a number of daily cues called *zeitgebers* (German for "time givers"). You can think of this process as being akin to a clock that keeps fairly accurate time across 24 hours – sometimes it's a little slow, sometimes a bit too fast – but never very far off since it is reset every day by the appearance of cues such as light/dark cycles, availability of food, social interaction, environmental noise, and so on.

The study of biological rhythms has become one of the hottest topics in science in the last few years, particularly since they are often altered in disease states. It is believed by many investigators that a better understanding of the mechanisms generating these rhythms may also shed light on the disease processes that affect them.

Every year the prestigious journal *Science* lists the ten most important discoveries of the year across all fields. For two years in a row (1997 and 1998) the study of the genes responsible for regulating the body's clocks has made the Runners Up list.

> "Nineteenth-century philosophers proposed that God was a clockmaker who created the world and let it run. Modern biologists might in part agree, for it's clear that evolution has crafted clocks that allow almost all organisms to follow the rhythm of the sun. In 1998, a volley of rapid-fire discoveries revealed the stunning universality of the clock workings. Across the tree of life, from bacteria to humans, clocks use oscillating levels of proteins in feedback loops to keep time. Perhaps more amazing, fruit flies and mice – separated by nearly 700 million years of evolution – share the very same timekeeping proteins. Now that they better understand the cellular clock, scientists can begin to manipulate it, with applications from curing jet lag to brightening winter depression". [*Science*, December, 1998]

Although it is difficult to find an environment completely lacking of external cues that might be used to reset timing, such experiments have shown that circadian rhythms persist even in their absence. That is, these rhythms maintain an approximately 24-hour period, but since their timing is not exactly 24 hours, they come in and out of synchrony with astronomical time.

An example is given in Figure 13-1, which shows the daily plot of a person's sleep-wake cycles. Each horizontal line is a day with solid lines indicating sleep and broken lines indicating waking. The triangles indicate the point of the day's lowest body temperature.

In this case, the subject first encountered 9 days of natural 24-hour cycles of light and dark, noise and quiet, and temperature changes. All timing cues were removed during the next 25 days, however, and the subject was free to set his own schedule. Two things become apparent from the data. First, without timing cues, the subject's "daily" schedule lengthens to about 25 hours, which becomes very stable from day-to-day. Second, since the new period is 25 hours, the subject comes in and out of synchrony with the natural 24-hour cycle. This is known as free-running behavior.

An interesting aspect of this behavior is that the low point of body temperature, which usually occurs just before waking, shifts to the beginning of the sleep period in the free-running condition. Recent studies have demonstrated that many physiological variables such as core body temperature and others change reliably over a 24-cycle, even in individuals that have a longer or shorter sleep-wake cycle. This means different clocks must regulate these variables, and that they can become desynchronized under certain conditions. (For example, if body temperature is operating on a 24-hour period, but sleep-waking cycles change every 25 hours.)

The consequence of desynchronizing the different clocks regulating these variables is currently a matter of considerable debate. Body temperature, for instance, regulates a number of physiological processes that might critically depend on slower biochemical reactions at the point in time just before waking. During the first 9 days, this would work just fine in our example. However, in the free-running condition, the relative timing between core body temperature and sleep-wake cycles becomes unstable and shifts to the beginning of a sleep session.

The ability of the clocks to desynchronize has led some to suggest that such a process may underlie certain disease conditions, including seasonal affective disorder and major depression. All is not hopeless, however, in that once the timing cues are re-introduced into the environment (at day 35 in Figure 13-1), the timing between core body temperature and sleep-wake cycles returns to normal values. In other words, *the presence of zeitgebers is important both for resetting individual clocks to a 24-hour period and for synchronizing different clocks that control independent processes.*

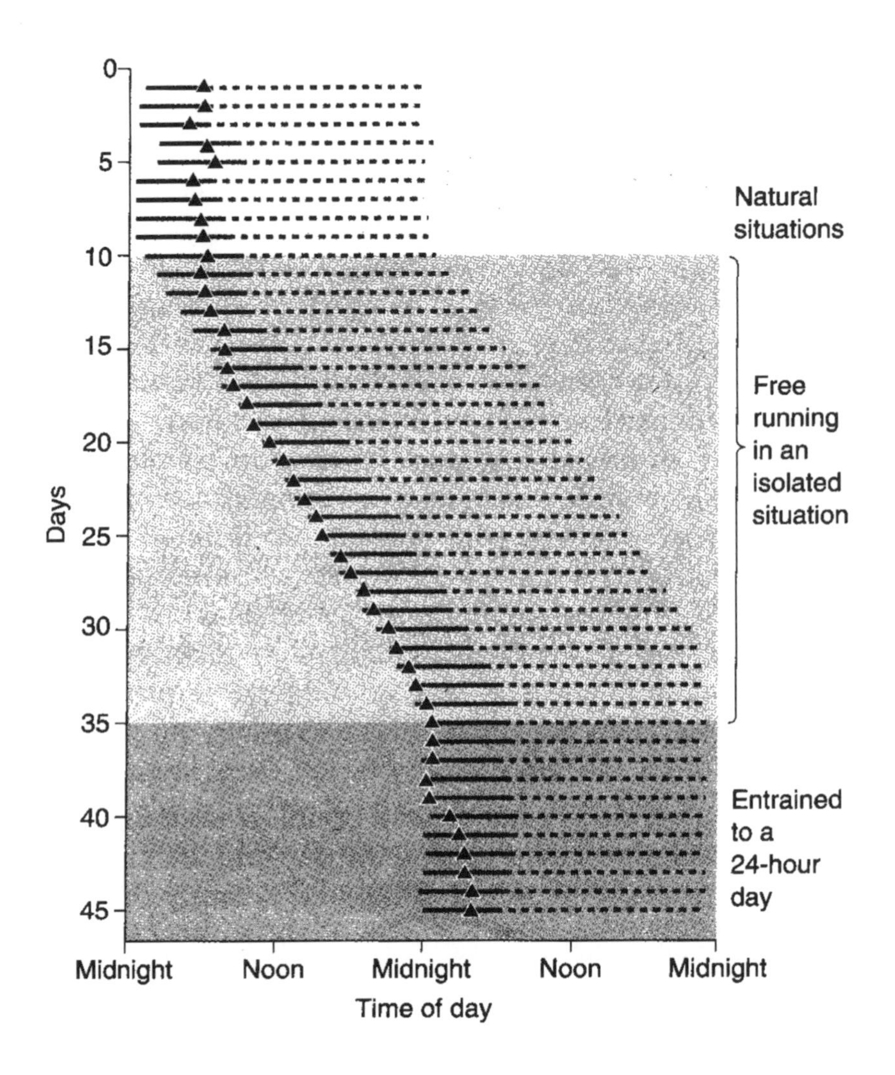

Figure 13-1 – The sleep-wake profile of a single subject across 45 days. Each horizontal line indicates a day. Solid lines indicate sleep, while dashed lines correspond to waking. The triangle marks the point of the day where the lowest core body temperature was recorded. During the first 9 days, the subject encountered natural 24-hour cycles of light and dark along with changes in environmental noise. During this period, the subject exhibited a fairly

stable sleep-wake cycle of approximately 24 hours with body temperature at its minimum typically toward the end of sleeping. During the middle 25 days, however, all timing cues were removed, and the subject was free to set his own schedule. As can be seen from the figure, the subject developed an approximately 25.5 daily cycle that gradually shifted with respect to the 24-hour period of the astronomical day. This loss of entrainment to the natural 24-hour period is known as "free-running". Also note that the minimum core body temperature now shifts to the beginning of the sleep period, which may cause disruptions to multiple temperature-dependent processes that are circadian in nature. On days 35 through 45, the timing cues are brought back into the environment and the subject gradually returns to a stable 24-hour rhythm with the minimum body temperature appearing late in the sleep cycle. This experiment indicates the importance of being able to utilize external cues to reset different circadian rhythms so they remain in a stable relationship to one and other.

The Hypothalamus: A Brain Clock

A biological clock that regulates circadian rhythms has essentially three components: an input pathway, the clock, and an output pathway. Let's continue with our example given in Figure 13-1.

The major zeitgeber that resets our sleep-wake cycle is the dark-light cycle that accompanies the rotation of the earth. Photoreceptor cells in the retina are sensitive to light and send this information to several downstream brain regions. One such region is a tiny cluster of cells in the hypothalamus called the suprachiasmatic nucleus (SCN – Figure 13-2). Its name comes from the fact that the SCN is located just above the optic chiasm where the optic nerves from both eyes cross.

Cells in the SCN oscillate on their own natural frequency of about 24 hours. You can harvest these cells and grow them in a petri dish - dissociating them from the rest of the brain and any light-dark cues - and they will still display changes in excitability that vary uniformly over a

24-hour period. The cells have thus evolved a set of intrinsic biochemical reactions that cause them to vary their firing rate at roughly the same period as an astronomical day. Pretty cool, huh?

If you remove the SCN from laboratory animals, they no longer display 24-hour periods to many typically circadian phenomena such as body temperature fluctuations, sleep-wake cycles, and hormonal secretion. In one example, a normal squirrel monkey is kept in a constantly lit environment, resulting in an intrinsic rhythm of about 25.5 hours in sleep-wake behavior and core body temperature. The activity states were defined as awake, slow-wave sleep and rapid eye movement sleep (when dreaming occurs). This rhythmicity was completely abolished in monkeys that were kept in the same environment, but had the cells of their SCN removed. If SCN cells are then harvested from a donor and transplanted into the lesioned animal, rhythmicity is restored to these behaviors but seems to reflect the new period of the donor rather than the host. Thus it appears that each SCN cell is like a mini clock with a preference for approximately 24-hour periods.

Moreover, this period can be advanced or delayed by changes in the natural light-dark cycle. For example, creating a condition that leads to light appearing 2 hours earlier than normal will similarly advance multiple variables that are regulated according to a circadian rhythm by about 2 hours.

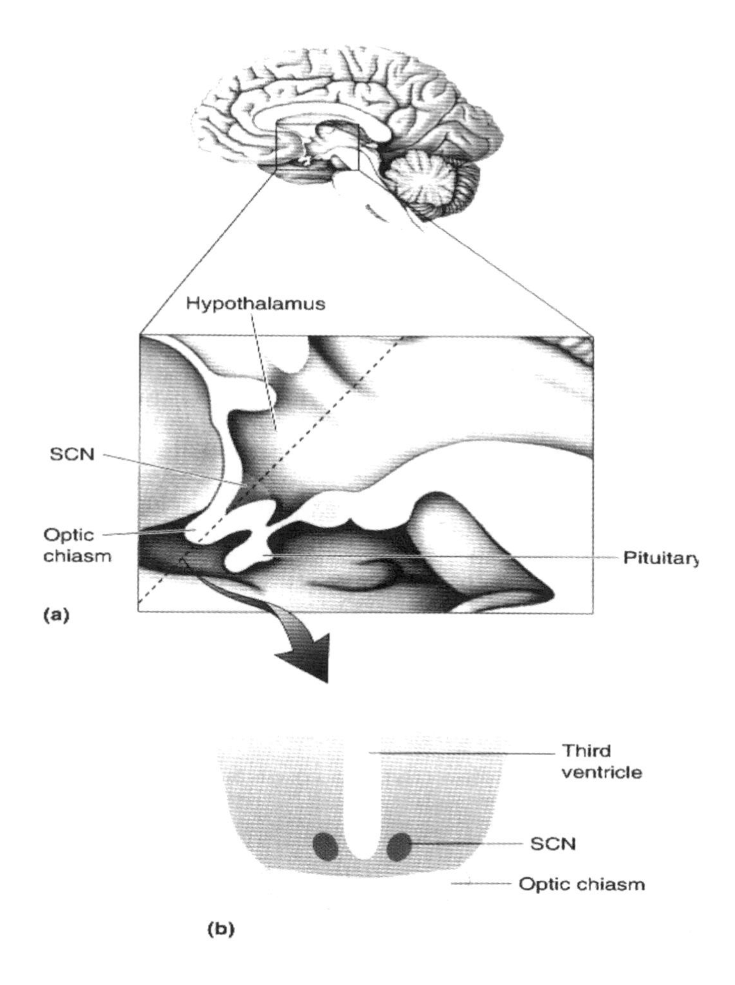

Figure 13-2 – A cross section cut through the center of a human brain (the front is on the left). In (a) the boxed region is expanded to show several components of the hypothalamus and surrounding area. Another section, this time taken at an oblique frontal angle (dotted line in (a)) reveals (b) the two suprachiasmatic nuclei (SCN) located on the left and right sides of the brain just above the optic chiasm (see text for details). [Reproduced with permission from Bear, Connors, and Paradiso, 2001,

Serotonin regulates the synchrony of circadian rhythms

We now know the SCN is an incredibly complicated region, utilizing multiple transmitters and peptides to communicate with other brain regions. A robust connection from the raphe nucleus of the midbrain to the SCN results in serotonin being released into the SCN at times when raphe cells are most active (i.e. during awake, alert behaviors).

Raphe cells are less active, and consequently release less serotonin, during periods of sleep or drowsiness. This has important implications. It turns out that serotonin reduces the ability of SCN cells to respond to light stimulation of the retina. In other words, during peak waking hours when the SCN receives maximal serotonin, it is less able to fire in response to photoreceptor stimulation by daylight. As the day continues toward twilight, however, the raphe neurons begin to quiet down, release less serotonin into the SCN, and thus release it from this blockade. Since there is still light at this time, the SCN cells begin to fire vigorously between twilight and dusk. When they fire, they release both the neurotransmitter GABA (that usually inhibits target cells) and a neuropeptide called vasopressin that has multiple functions in the brain and endocrine system.

Vasopressin-containing neurons of the SCN project to the periventricular region of the hypothalamus (Chapters 3-5) and inhibit these cells from releasing corticotropin releasing factor (CRF). Thus once cells of the SCN become activated, they inhibit the release of CRF from the hypothalamus, which in turn reduces activity in the HPA axis.

Consequently, in addition to regulating those processes we typically think of as being circadian such as sleep-wake cycles, changes in metabolic rate and core body temperature, the periodicity of SCN cell activity also influences HPA axis and immune function (Figure 13-3). This property may help us understand why circadian rhythm abnormalities are often associated with mood disorders. Furthermore, it may lead to novel therapeutic interventions for the subset of depressives that exhibit this "circadian sensitivity".

Circadian Rhythms and Mood

Disturbances of sleep are typical for most patients suffering from major depression and are considered to be one of the core symptoms in diagnostic tools such as the DSM. It has been reported that over 90% of depressives experience some type of sleep disturbance.

Typically patients have trouble falling asleep, frequent nocturnal awakenings, and early morning awakenings. The latter symptom is a classic signature of endogenous depressive disorder. Thus insomnia is commonly associated with major depressive disorder. Hypersomnia is less typical of depressives, but is often observed in some subtypes of depression, particularly bipolar disorder.

In addition to the behavioral sleep disturbances, the architecture of sleep is also altered in depressed patients as can be seen in scalp recorded brain EEG (electroencephalogram) patterns. Depressives typically exhibit a decrease in slow wave sleep, dominated by low frequency, large amplitude EEG waves, and an increase in REM sleep (dream sleep), dominated by high frequency, small amplitude waves.

So what does all this mean and why should I care? It turns out that we have a reasonable understanding of the neurobiological mechanisms that drive each stage of sleep, resulting in these particular wave patterns. They are controlled, in part, by changes in the firing patterns of cell groups in the midbrain that release serotonin, norepinephrine and acetylcholine into neocortical structures at different times of the day and during different stages of sleep.

For instance, these groups directly regulate the appearance of REM sleep. Periods of REM sleep are associated with a sudden increase in the activity of acetylcholine-containing cells of the pons (a part of the midbrain). The activation of these cells results in acetylcholine being released into multiple cortical brain regions that sets the stage for REM sleep to occur.

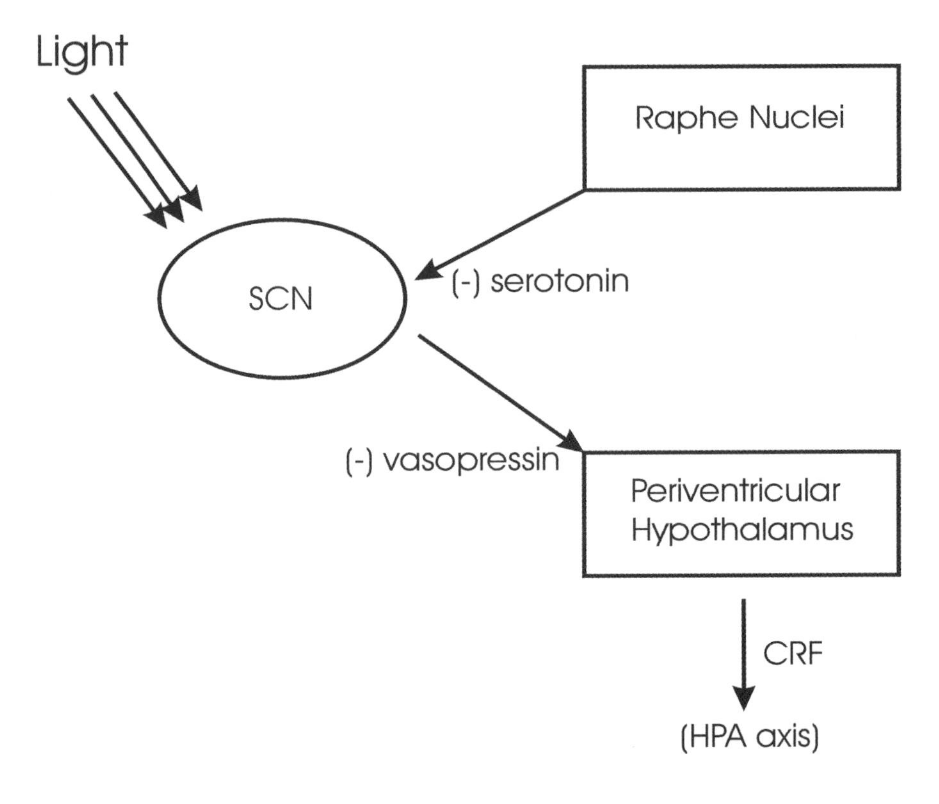

Figure 13-3 – Light normally stimulates activity of SCN cells, but this activation is inhibited during peak waking hours by serotonin, released from the raphe nuclei. Cells in the raphe nuclei become less active during evening hours and release less serotonin into the SCN as a result. At twilight, when there is still enough light to stimulate the SCN, but less inhibition from the raphe nuclei, cells in the SCN become active and release the neuropeptide vasopressin. Vasopressin, in turn, inhibits the release of corticotropin releasing factor (CRF) into the HPA axis. Thus, the 24-hour circadian rhythm that is observed in SCN cell firing has a direct impact on the extent to which the HPA axis is activated. This may have important consequences for understanding why circadian processes are not only affected during depression, but may also contribute to the development of the illness.

REM sleep is turned off by a subsequent increase in serotonin and norepinephrine cell activity in the raphe nuclei and locus coeruleus respectively. Thus, the monoamine systems that are suspected to play a role in the etiology of mood and its disorders are also critical for sleep regulation, and as it turns out many other circadian processes.

An exciting prospect is that the relationship between mood and sleep disorders is not simply one-way with depression causing insomnia, but is rather bi-directional. Recent studies have suggested that insomnia may be an independent risk factor for the subsequent development of major depression.

Epidemiological data obtained from general practitioners by Mathias Berger and Dieter Riemann at the University Hospital of Freiburg, Germany have shown that the likelihood of major depression is increased four-fold in patients with severe insomnia[1]. Comparable results have been observed in the United States.

This means that either these folks are depressed and thus have trouble sleeping, or that the sleep disturbances result in a type of "learned helplessness" that precipitates depressive illness. Consistent with the second interpretation, studies have recently shown that sleep disturbances often *precede* other symptoms specific to major depression such as mood abnormalities. This can be thought of as being akin to an early warning system.

Clock genes

A third possibility is that neither the sleep nor mood disturbances are causal to the other, but are rather a consequence of some disrupted mechanism further upstream whose output influences both of these functions. A candidate for this role is the set of genes and their associated transcription regulators (see Chapter 1) that control the timing of different biological rhythms and their synchronization with each other.

Several genes that code for proteins involved in timing have now been identified and given fanciful names like *period*, *frequency*, *cycle*, *doubletime*, *time*, *clock*, and so on. Although the details of these molecular/genetic clocks vary from species to species, the general mechanism at work was discovered in a series of recent experiments performed in the laboratory of biologist Joseph Takahashi at Northwestern University. His team showed that timing is established by periodic changes in gene expression patterns that work through a simple negative feedback loop[2].

A clock gene is first transcribed to produce a messenger RNA strand that is then translated into proteins. After a specific delay, the new proteins are phosphorylated (a phosphate group is added) that results in them inhibiting any further gene transcription. Consequently, less transcription leads to less translation of new proteins, a process that reduces the likelihood of continued inhibition of transcription (Figure 13-4).

Through this negative feedback loop, a clock emerges based on rising and falling levels of gene expression patterns and resultant protein levels. The period of the clock is dictated by the delay involved in the phosphorylation process and the end protein products of gene transcription can directly alter the excitability of the cell and how it responds to stimulation. In the SCN, the cycle of gene transcription repeats approximately every 24 hours, resulting in the same pattern of SCN cell activity changes.

It is quite possible that a disruption of this molecular clock based on gene expression may contribute to abnormalities in one or more processes that are normally regulated by circadian rhythms. This disruption will lead to inconsistent timekeeping, or a difficulty in resetting the clock to the daily changes in light-dark cycle.

Following this logic, many of the symptoms seen in mood disorders that are either directly or indirectly influenced by circadian rhythms (e.g. sleep, eating, drinking, various hormonal abnormalities) may result from a desynchronization of these normally synchronized rhythmic processes. This suggests treatment therapies that target the gene expression regulators themselves may be effective antidepressants. These include drug interventions, or behavioral treatments such as light therapy, which stimulates gene expression in the SCN.

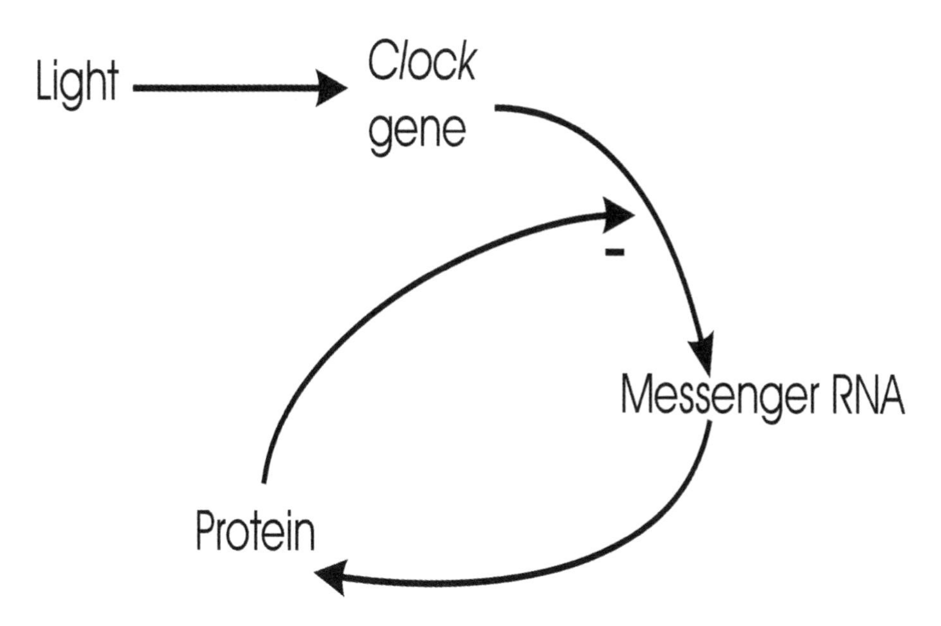

Figure 13-4 – Light activates photoreceptor cells in the retina that pass this information on through a series of additional stops en route to the SCN located in the hypothalamus. This activation induces the transcription of the *clock* gene into a messenger RNA transcript. The messenger RNA transcript is then translated into a protein (see Box 2-1) that, after a delay period, inhibits the further transcription of the gene. This delayed negative feedback loop results in a periodic cycle of gene expression that subsequently generates proteins that regulate the excitability of the cell. In this manner, SCN cells exhibit rhythmic firing with a delay period of approximately 24 hours.

Some investigators have suggested that behavioral treatments that help stabilize and/or synchronize different circadian rhythms that are out of synchrony (such as sleep-wake cycles and changes in cortisol production), will also have beneficial effects on depressive illness. The reasoning being that if depression is caused by these rhythms becoming destabilized and irregular with respect to one and other, why not try to treat the disorder with behavioral therapies that rectify this specific problem.

There are currently several groups investigating this hypothesis, employing a variety of different treatment strategies depending on the particular rhythmic process being considered. Among the more popular treatment approaches are different combinations of the following: light therapy, sleep-wake cycle advance or delay; meditation and other forms of stress reduction; and biofeedback to name just a few.

The hypothesis implies that in combined therapies, timing is critically important. So just getting light therapy at any old time of the day is not expected to be helpful. Rather treatments such as these should be administered carefully, at precise times, as advised by a knowledgeable physician or clinician.

Chapter 14

†

New Vistas in our Understanding of Mood

As we begin the final chapter in this story, hopefully several points are now clear to readers:

- Mood disorders are complex syndromes that can often involve disruptions in multiple physiological systems (e.g. immune response, stress reactions, changes in monoamine transmission, changes in gene expression, etc.).

- The etiology of mood is also complex and involves the interaction of genetic, physiological, psychological, and environmental variables. Moreover, it is misleading to think these variables are easily separable into independent components, each with a unique contribution to the development of pathological conditions.

- Although these interactions create extra challenges to our understanding the biology of mood and its disorders, they also raise new possibilities for developing novel treatment approaches that target these systems. These may come in the form of drugs, behavioral modification strategies, or the emergence of new forms of psychotherapies.

Let us now consider some open questions that we only touched on in the preceding chapters. The first concerns learning.

Learning and Plasticity Revisited

A recent book on affective disorders suggests that clinical depression does not have a root biological cause, but rather is the result of learning maladaptive patterns of thinking and behavior. Most of the examples given were based on the "simple" mechanisms of classical and operant conditioning that we discussed in Chapter 6.

For those of us working in the field of plasticity, it has become increasing clear that the biological underpinnings of learning and memory are incredibly complex. We are just beginning to understand the genetic, molecular, and cellular mechanisms that provide the functional hardware for you to remember a new phone number, that fire is hot, and what a cytokine is.

Furthermore, the plasticity that supports these functions operates on a number of levels. We all know that learning can be demonstrated at the behavioral level, yet most of us probably do not appreciate the fact that there are robust changes that occur at the molecular level inside cells, which form the physical basis for this memory. Interestingly, although different cellular and neurotransmitter types are involved in learning and memory in different species, the molecular and genetic components seem to have a great deal in common – from fruit flies to primates.

Nobel prize-winning neuroscientist Eric Kandel of Columbia University was one of the first to demonstrate that comparable molecular mechanisms and subsequent changes in gene expression patterns are involved in memory formation in different species. In other words, there seems to be a final common pathway that supports the function. This is another example of the way biological mechanisms are evolutionarily conserved across species.

Many of the brain regions (e.g. hippocampus, amygdala, prefrontal cortex) and transmitter/receptor proteins (e.g. NMDA glutamate receptors, acetylcholine, BDNF, glucocorticoids) that are known to be critically involved in learning and memory have recently been implicated in the regulation of mood. This suggests that some of the anatomical changes that accompany mood disorders such as brain tissue reduction and changes in metabolism/glucose utilization, may share molecular/genetic features with other forms of plasticity.

Animal models have shown that the reversal or "normalization" of many of these changes depends on the activation of the same neurotransmitter receptors and intracellular messengers (e.g. cyclic adenosine monophosphate - cAMP) that are required for normal learning and memory.

It is tempting, then, to suggest that a final common pathway in the etiology of mood disorder may involve a similar set of molecular chain reactions and resultant changes in gene expression. These changes could lead to aberrant protein synthesis and ultimately pathological symptoms. Moreover, the same process may also be involved in the correction of these symptoms with successful antidepressant treatment using drugs and/or psychotherapy.

This might explain why such treatments take weeks to work (since they would actually require protein synthesis) even though their cellular effects are immediate. It might also explain why so many different forms of treatment seem to have antidepressant properties even though they are known to target widely different brain regions and transmitter families. In each case, perhaps the treatment is setting up the prerequisite changes that eventually lead (at least with successful treatment) to the necessary molecular/genetic alterations that *actually* foster recovery.

Cyclic AMP, CREB, and Antidepressants

There is now a large body of evidence that suggests monoamine transmitter systems are involved in major depressive disorder. The big question, of course, is *how* are they involved? As was discussed in the introduction, the strongest evidence for the "monoamine theory of depression" comes from the fact that many effective antidepressants are believed to increase the concentrations of these transmitters in the synaptic cleft (for instance by blocking reuptake or the enzymes that metabolize them).

Experiments that involve the depletion of certain monoamine transmitters and examine subsequent changes in mood have had less consistent results. If, as the theory goes, clinical depression is caused by a deficit in monoamine transmission, then a depletion of these transmitters induced by dietary restrictions or pharmacological manipulations should produce the disorder, right?

Well, not so fast. It turns out that most healthy subjects do *not* become depressed with such manipulations. However, patients who experienced remission from depression by using either SSRIs or SNRIs, do indeed relapse following these procedures. These results suggest that monoamine transmission may be particularly involved in the *maintenance* of an antidepressant response.

We also learned that some very good antidepressant treatments do not seem to target monoamine transmission at all. For instance, popular new medications like the aminoketone, bupropion (Wellbutrin), as well as several CRF antagonists that are currently showing encouraging results in clinical trials have little or no direct effects on monoamine transmission. Thus we have some effective antidepressant medications that target serotonin, others that target norepinephrine, or acetylcholine, or CRF, etc.

So, what is the common denominator? If you want to ask the same question, but sound more like a biologist, you would query, "what is the final common pathway"? The assumption being that all of these diverse treatment strategies eventually result in the same set of changes, perhaps at the molecular level, that are the primary force driving the alleviation of mood disorder symptoms.

Let's consider a candidate pathway recently proposed by scientists Ronald Duman, George Heninger, and Eric Nestler of the Yale University School of Medicine[1]. Many early studies demonstrated that long-term antidepressant treatments (generally 3 weeks or longer) reduce the number of receptor sites for serotonin and norepinephrine. The most common examples of this are observations of decreased beta-adrenergic receptors (a subtype of norepinephrine receptor) and serotonin-2 receptors in the limbic systems of laboratory animals treated with tricyclics or SSRIs.

A common point between these two transmitter subtypes and other systems that support antidepressant actions is the induction of a series of chemical reactions inside the cell that begins when the transmitter is bound to the receptor. We call this a *signal transduction pathway* because it is involved in transducing the incoming chemical messages that a cell receives in the form of different transmitters, into an electrical output message to postsynaptic cells located downstream.

One signal transduction pathway that is altered in depressed patients is the cyclic AMP (cAMP) – CREB cascade that occurs when a transmitter binds to a host receptor. Let's look at an example illustrated in Figure 14-1.

When serotonin or norepinephrine binds to a receptor on the cell surface, this causes the activation of an associated G-protein that, indirectly (the numerous intervening biochemical reactions omitted for purposes of sanity retention) results in the production of the second-messenger signaling molecule, cAMP. Cyclic AMP is referred to as a "second messenger" because the primary messenger is the neurotransmitter binding to the receptor.

Increased production of cAMP inside the cell leads to a number of subsequent reactions that include the production of protein kinases (little enzymes that add phosphate groups to other proteins that speed up reactions) and CREB (cAMP-response-element-binding protein).

CREB is actually a gene transcription factor that, once phosphorylated, regulates the expression of other end-effector proteins that have multiple purposes inside and outside the cell.

One such end product protein is BDNF (brain-derived neurotrophic factor). As you learned earlier, BDNF is a neurotrophic factor, meaning it is involved in the growth and maintenance of brain cells. We mentioned some of the effects of this substance in Chapter 5 (Birth and Death of a Neuron). BDNF plays a role in neuron survival and, it is currently thought, the genesis of new brain cells.

Recent experiments in my laboratory have also shown that BDNF facilitates a type of plasticity in adult cells that is thought to underlie learning and memory called long-term potentiation (LTP). Thus BDNF seems important for a number of trophic actions that result in the increased function and survival of existing brain cells as well as the birth of new ones.

Activation of CREB proteins most likely represents a common therapeutic target for different antidepressant treatments. Research has now shown that there are subtypes from each monoamine transmitter system that activate the cAMP-CREB pathway, and that this reaction cascade is activated with long-term antidepressant treatments.

The Yale group has demonstrated an increase in CREB and associated messenger RNA (a marker for CREB production) in the hippocampus of rats that underwent long-term treatment with SSRIs and SNRIs[2]. A very exciting point is that the time course for the induction of

CREB following antidepressant treatment is approximately the same as that observed for the therapeutic effect of these agents to begin alleviating depressive symptoms (about 2-3 weeks).

Remember back in Chapter 10 we mentioned that depressed patients exhibit hippocampus abnormalities including changes in glucose metabolism, regional cerebral blood flow, and tissue volume. It is tempting to speculate that these patients may have an abnormally low activation of the cAMP-CREB pathway, resulting in diminished concentrations of BDNF and its protective benefits to this region.

Successful antidepressant treatments may have a common element in that they all induce changes that activate this pathway and the subsequent production of BDNF and other end-product proteins necessary for the proper functioning of limbic brain cells.

Consistent with this interpretation, stress (both physical and psychosocial) can often trigger a major depressive episode, particularly in those with a pattern of recurrence. It is well known that stress also produces a reduction in BDNF in a number of neocortical and limbic brain structures. One can imagine, then, a *cumulative* effect of stressors that cause some degree of limbic system damage (for instance to cells of the hippocampus), making the region more susceptible to future trauma. Subsequent reductions in BDNF, whether due to stress or alterations in the cAMP-CREB signal pathway, might then result in even greater damage to the region and symptoms of enhanced severity.

This process may also help explain why other illnesses associated with neuronal damage often predispose an individual to mood disorders. Examples of these include hypoxia-ischemia, hypoglycemia, stroke, neurotoxic damage, and a number of viral infections (also see the Chapters 9 and 10 on immune function).

These observations suggest that therapeutic treatments for mood disorders may be found in chemical compounds or behavioral approaches that facilitate the cAMP-CREB pathway or the production of BDNF in limbic structures. Several groups have now demonstrated that BDNF has antidepressant effects in two animal models of depression, the forced-swim (acute stress) and learned helplessness paradigms.

Similarly, long-term antidepressant treatment with tricyclics or SSRIs has been shown to increase BDNF expression in the hippocampus of laboratory animals. Indeed, the Yale group has also shown that long-term antidepressant pretreatment blocks the typical reduction of limbic system BDNF that is normally induced by stress. These results are summarized in Figure 14-2.

One intriguing therapeutic prospect of these findings is that classical antidepressant treatments (e.g. SSRIs) combined with compounds that facilitate BDNF or prevent it from down-regulating due to stress (i.e. by using a CRF blocker), may have even greater efficacy and speed of onset in ameliorating depressive symptoms than either strategy alone. This is because the signal transduction pathway is targeted directly in addition to mechanisms upstream that will *eventually* result in its modification, such as by enhancing monoamine transmission. At present, several research groups are testing this hypothesis using phosphodiesterase inhibitors (e.g. rolipram), which prevent the normal metabolic breakdown of cAMP, and consequently promote CREB production.

Pharmacogenetics and Pharmacogenomics: The Future?

The treatment and diagnosis of mood disorders has undergone many changes in the last decade. We have certainly adopted new approaches and therapies, but I would argue that a fundamental understanding of the biology of mood and its disorders has not been forthcoming. This has led to "hit or miss" drug therapy.

There are four general classes of antidepressants in use at present (see Appendix I for details), the tricyclics, MAOIs, SSRIs, and the "atypical" compounds, the latter being the "everything else" category. Most often the hard work really begins once a diagnosis of major affective disorder is made and both patient and physician agree that pharmacological treatment should be considered. The patient will be prescribed an antidepressant drug that, in many if not most cases, is unfortunately based on a rather arbitrary decision. Indeed, tolerance of side effects is the overwhelming variable guiding the prescription of antidepressants today. If Drug A doesn't begin to work after 4-6 weeks, either the dosage will increased or another drug will be prescribed (again rather arbitrarily), often from another class.

This process can take months, and remember we're talking about people often in extreme pain and perhaps even suicidal. This feature also makes clinical studies very difficult, which are plagued by high dropout rates and morbidity (and sometimes mortality). How can we improve this situation?

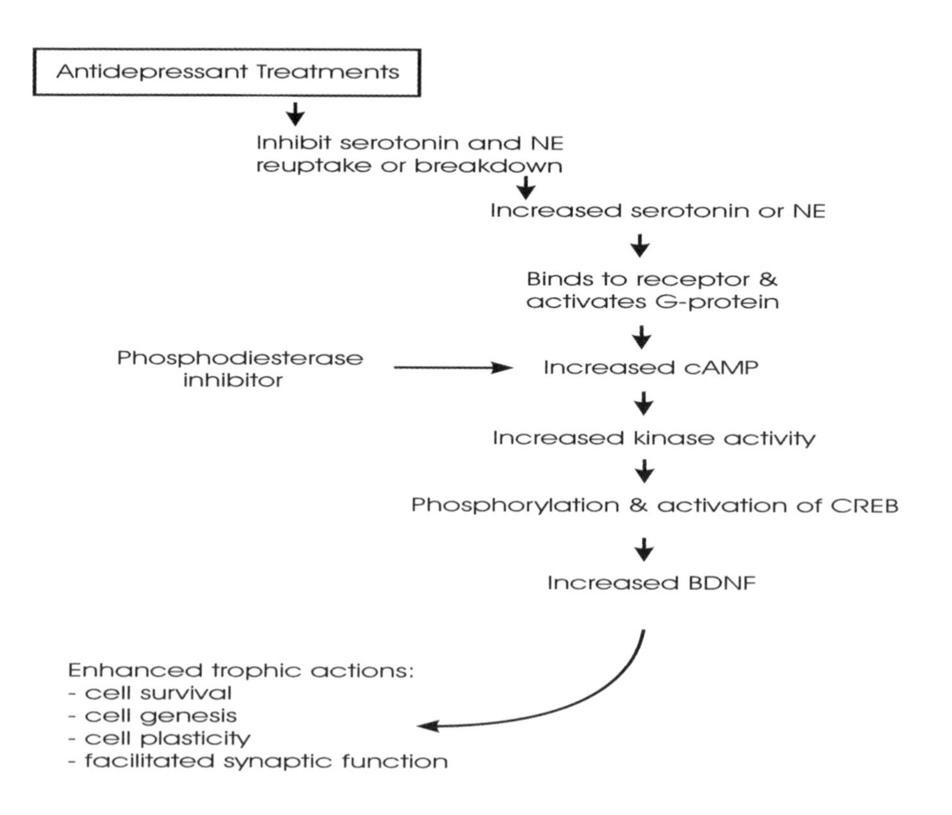

Figure 14-1 – The cAMP-CREB-BDNF pathway. Antidepressant treatment can lead to an elevation of either serotonin or norepinephrine in the synaptic cleft. This increases the likelihood that neurotransmitter will bind to and activate a postsynaptic receptor. This, in turn, triggers a cascade of reactions starting with the induction of a G-protein that mediates increased cAMP and protein kinase activity. Protein kinases phosphorylate (add phosphate groups to) CREB gene transcription factors located in the nucleus of the cell, resulting in the expression of genes that code a variety of end-product proteins such as brain-derived neurotrophic factor (BDNF). BDNF has been shown to have several trophic actions in limbic structures such as the hippocampus that include facilitating brain cell genesis, survival, and plasticity. See text for additional details.

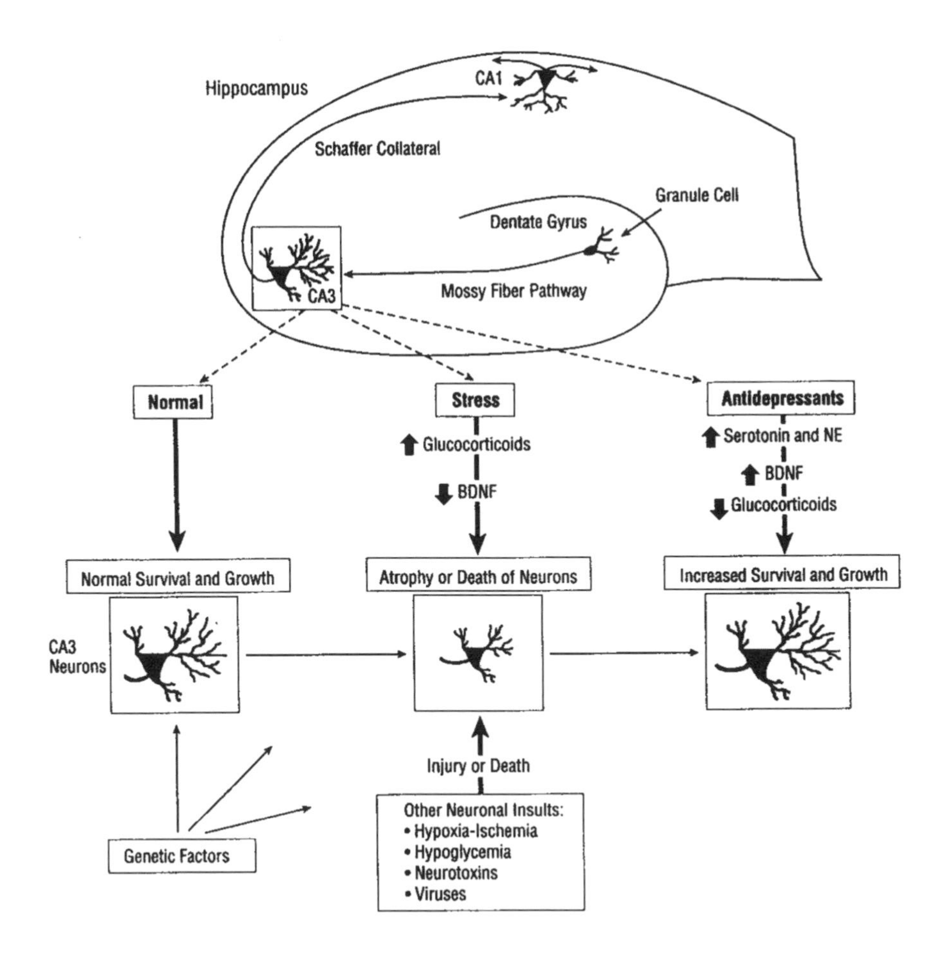

Figure 14-2 – The stress-induced production of glucocorticoids can result in a decrease of the trophic factor BDNF. Decreased BDNF has a negative impact on hippocampus cell survival, genesis, and plasticity. Other variables including genetic factors and a number of disease conditions may exacerbate neuronal injury or death. Together these variables and the stress-induced decrements in BDNF will play a causal role in the changes in hippocampus metabolism and volume seen in many depressed patients. Long-term antidepressant treatment has been shown in laboratory animals to increase BDNF, while decreasing glucocorticoids and

depressive symptoms in both the acute stress and learned helplessness models of major depressive disorder. Such treatments also prevent the stress-induced changes in hippocampus volume typically observed in these models. [Reproduced with permission from Duman et al. (1997) *Archives of General Psychiatry*, 54, 597-606.]

One possibility can be found in the convergence of the two fields, genetics and genomics. The aim is to develop genetic markers that *predict pharmacological or behavioral/psychotherapeutic* treatment responses, thus potentially saving time, expense, and pain to the patient. One strategy being used by several investigators is the search for genetic polymorphisms. Polymorphisms are genotypes that are repeated within a cell population with an exaggerated frequency that cannot be explained by recurrent mutation rates alone.

A prime example of this is the search for polymorphisms that associate antidepressant treatment response with changes in the expression patterns of the serotonin transporter gene. This transporter protein, as you'll recall, is the target for selective serotonin reuptake inhibitors (SSRIs) like Prozac.

Recently, a functional polymorphism was identified on the regulatory region (see Chapter 1) of the serotonin transporter gene that, as you have probably guessed, modulates the transcription of the gene. Incredibly, this polymorphism has been associated with the antidepressant effects of both SSRI and short-term sleep deprivation treatments. Thus, *drug and non-pharmacological treatments may share the same genetic/molecular mechanism responsible for their antidepressant actions.*

The applicability of this finding is truly impressive. It suggests that a prescreening of this gene polymorphism can be used to predict whether or not a patient will respond to these classes of treatments, saving time, minimizing patient distress, and perhaps avoiding mortality.

New genomic tools such as "Total Gene Expression Analysis" and high-speed PCR (polymerase chain reaction) driven by improved microarray processing technologies are now being applied to study the genomic effects of successful antidepressant treatments. This search is driven by the assumption that different treatments acting on diverse primary pathways, may ultimately owe their therapeutic efficacy to adjustments made to common downstream mechanisms (e.g. cAMP-CREB pathway or BDNF production). Indeed, these tools were used to first identify new gene transcripts that were equally regulated by chronic treatment of antidepressants from completely different classes, in this case imipramine (a tricyclic) and flouxetine (an SSRI).

The long-term goal then, is to develop genetic markers that can be used by the physician to determine the best treatment strategy for the individual. When one considers the incredible expense, both personal and to society as a whole, associated with "hit-or-miss" treatment and diagnostics, it would seem that a significant investment in additional genetic and genomic studies of mood and its disorders at the federal level is warranted. Indeed such research will no doubt be of enormous scientific, clinical, and public health value.

Genetic markers could also complement classical diagnostic tools such as the DSM, which at present does not have even a single physiological marker or assay included in its definition criteria for major affective disorders.

I leave you, my readers, with a set of appendices for additional information and the sincere hope that you have benefited in some way from this book. I welcome any comments you may have, small or large, and wish you health, happiness, and peace.

Appendix I

†

Current Antidepressant Drugs

Selective Serotonin Reuptake Inhibitors (SSRIs)

Drug	Predominant Action	Major Side Effects
Citalopram	Serotonin	GI discomfort
Fluoxetine	Serotonin	GI discomfort, sexual dysfunction
Fluvoxamine	Serotonin	GI discomfort
Paroxetine	Serotonin	GI discomfort, sexual dysfunction
Sertraline	Serotonin	GI discomfort, sexual dysfunction

Selective Norepinephrine Reuptake Inhibitors (SNRIs)

Drug	Predominant Action	Major Side Effects
Maprotiline	Norepinephrine	Insomnia
Reboxetine	Norepinephrine	Insomnia

Monoamine Oxydase Inhibitors (MAOIs)

Drug	Predominant Action	Major Side Effects
Phenelzine	MAOI	Hypertension
Tranylcypromine	MAOI	Hypertension
Moclobemide	MAOI	Minimal

Tricyclic Antidepressants

Drug	Predominant Action	Major Side Effects
Amitriptyline	Serotonin/ norepinephrine	Sedation, dry mouth
Clomipramine	Serotonin	Dry mouth
Desipramine	Norepinephrine	Increased heart rate
Doxepin	Norepinephrine/ serotonin	Sedation, dry mouth
Imipramine	Norepinephrine/ serotonin	Decreased blood pressure
Lofepramine	Norepinephrine	Increased heart rate
Nortriptyline	Norepinephrine/ serotonin	Dry mouth
Protriptyline	Norepinephrine	Insomnia
Tianeptine	Serotonin	GI discomfort
Trimipramine	Norepinephrine/ serotonin/DA	Sedation

Atypical Antidepressants

Drug	Predominant Action	Major Side Effects
Amoxapine	Mixed	Movement stiffness
Bupropion	Dopamine	Insomnia
Mianserin	Serotonin	Minimal
Nefazadone	Mixed	Headache
Trazadone	Mixed	Sedation, sexual dysfunction
Venlafaxine	Serotonin/ norepinephrine	Increased blood pressure

Appendix II

†

Resources for Patients and Researchers

There is now such an abundance of information "out there" on the internet, in journals, popular magazines, books, pamphlets, after-school specials, etc., that one needs to be clear that not all sources are created equal. There is good information, and, well, quite a bit of the other kind in circulation. Below I've listed a few sources that are generally considered reliable and informative.

www.nih.gov - I recommend you start your search here at the National Institutes of Health (NIH). From this location, you can link to a number of additional sites including the National Institute of Mental Health. They also have a depression hotline: 1-800-421-4211.

www.cdc.gov - The Centers for Disease Control and Prevention provides up to date information on a number of mental illnesses. Call 1-800-311-3435 for general information.

www.medscape.com - This is a wonderful resource used every day by folks in the medical profession. There is a very good psychiatry and mental health section.

www.fda.gov - The U.S. Department of Health and Human Services Food and Drug Administration is a good source of information about medications, their side effects, and current clinical trials. Food and Drug Administration 5600 Fishers Lane Rockville, Maryland 20857
1-888-INFO-FDA (1-888-463-6332)

www.veritasmedicine.com - This site has information about free trials and basic information on depression screening.

www.webmd.com - A good site for information and additional links on a diverse range of medical conditions and concerns.

Notes

†

INTRODUCTION – Autumn in Vermont

1. Harlow, J.M. (1868) Recovery from the passage of an iron bar through the head. *Publications of the Massachusetts Medical Society*, 2, 329-346.

Chapter 1 – Mood Genes

1. In *Neuroscience: Exploring the Brain*, M. Bear, P. Paradiso, 2000, Lippincott, Williams, and Wilkins.

Chapter 2 – Genes, Mood, and the Luck of the Draw

1. Gershon et al. (1982) A family study of schizoaffective, bipolar I, bipolar II, unipolar, and normal control probands. *Archives of General Psychiatry*, 39, 1157-1167; Maier et al. (1993) Continuity and discontinuity of affective disorders and schizophrenia: results of a controlled family study. *Archives of General Psychiatry*, 50, 871-883; Tsuang et al. (1980) Morbidity risks of schizophrenia and affective disorders among first-degree relatives of patients with schizophrenia, mania, depression, and

surgical conditions. *British Journal of Psychiatry*, 137, 497-504; Weissman et al. (1984) Psychiatric disorders in the relatives of probands with affective disorders: the Yale University-NIMH Collaborative Study. *Archives of General Psychiatry*, 41, 13-21; Weissman et al. (1993) The relationship between panic disorder and major depression: a new family study. *Archives of General Psychiatry*, 50, 767-780.

2. Sullivan et al. (2000) Genetic epidemiology of major depression: review and meta-analysis. *American Journal of Psychiatry*, 157, 1552-1562.
3. Kendler, KS (1995) Is seeking treatment for depression predicted by a history of depression in relatives? Implications for family studies of affective disorders. *Psychological Medicine*, 25, 807-814; Sullivan et al. (1996) Family history of depression in clinic and community samples. *Journal of Affective Disorders*, 40, 159-168.
4. Weissman et al. (1993) The relationship between panic disorder and major depression: a new family study. *Archives of General Psychiatry*, 50, 767-780.
5. Kendler KS (1998) Major depression and the environment: a psychiatric genetic perspective. *Pharmacopsychiatry*, 31, 5-9.
6. Hughs J (1986) Genetics of smoking: a brief review. *Behavioral Therapeutics*, 17, 335-345.
7. Fergusson DM & Horwood, LJ (1984) Vulnerability to life event exposure. *Psychological Medicine*, 14, 881-889.
8. Brown GW & Harris TO (1989) *Life Events and Illness*, Guilford Press, New York.
9. McGuffin et al. (1988) The Camberwell collaborative depression study: III. Depression and adversity in the relatives of depressed probands. *British Journal of Psychiatry*, 152, 775-782; Breslau et al. (1991) Traumatic events and posttraumatic stress disorder in an urban population of young adults. *Archives of General Psychiatry*, 48, 216-222; Plomin et al. (1990) Genetic influences on life events during the last half of the life span. *Psychology of Aging*, 5, 25-30; Lyons et al. (1993) Do genes influence exposure to trauma? A twin study of combat. *American Journal of Medical Genetics*, 48, 22-27.
10. Ogilvie et al. (1996) Polymorphism in serotonin transporter gene associated with susceptibility to major depression. *Lancet*, 347,

731-733.

11. Manki et al. (1996) Dopamine D2, D3 and D4 receptor and transporter gene polymorphisms and mood disorders. *Journal of Affective Disorders*, 40, 7-13.

Chapter 3 – stress, Stress, STRESS…!

1. Sapolsky, RM (1998) *Why Zebras don't get Ulcers*. W.H. Freeman and Company, New York.
2. This network of blood vessels is called the *hypothalamo-pituitary portal circulation*. In biology, we name a pathway connecting two regions by a combination of the path's origination and termination. This blood vessel portal begins at the hypothalamus and ends in the pituitary, hence the name "hypothalamo-pituitary portal".
3. Since the pituitary controls all these other glands, it was common to hear it referred to as the "master gland" in days past, but we now know that the hypothalamus is the real master gland of the endocrine system. When you encounter a stressor, cells in the periventricular region of the hypothalamus get excited and release CRF into the anterior lobe of the pituitary where it binds to specific receptors that only recognize its particular molecular shape. These receptors are located on the surface of anterior pituitary cells, and when CRF binds to them, the cells begin to secrete (or in some cases stop secreting) other hormones into general blood circulation that travel to all parts of the body. However, the hormones do not affect all regions, since only a small number of downstream glands (for instance the thyroid or adrenal glands) may have the specific receptors that match the hormone being released. So think of this business of hormone-receptor specificity kind of as a cell waiting for the secret password before beginning to activate targets sites in the rest of the circuit.
4. It should also be noted that the pituitary releases other hormones besides ACTH during the stress response. For example, it secretes prolactin, which suppresses reproduction during stress; both the pituitary and brain secrete endorphins and enkephalins, morphine-like chemicals that impair pain perception; vasopressin

which regulates your cardio-vascular response to stress; and several others.

5. As you can imagine, drugs that effect NE or ACh neurotransmission, can have a dramatic impact on ANS system activity. Drugs that either enhance NE neurotransmission or inhibit ACh neurotransmission are referred to as *sympathomimetic*, in that they cause effects that mimic sympathetic nervous system activation. This happens because a shift toward sympathetic activation will occur if this division is activated *or* if the parasympathetic division is attenuated. By contrast, drugs that inhibit NE transmission or enhance ACh actions in the periphery have a net effect that is *parasympathomimetic*. Examples of parasympathomimetics are drugs that block (act as antagonists of) the beta subtype of NE receptors. These so-called "beta-blockers" have the effect of slowing heart rate and lowering blood pressure – in other words, calming things down. Consequently, beta-blockers are often given to patients with high blood pressure, or as a mild sedative to prevent the physiological reactions that occur with stage fright. (I know of at least 3 fellow students who began taking beta-blockers in graduate school just before their dissertation defense. And, yes, they were calm as clams.)

Chapter 4 – Stress and the Brain

1. Heim et al. (2000) Pituitary-adrenal and autonomic responses to stress in women after sexual and physical abuse in childhood. *Journal of the American Medical Association*, 284, 592-597.
2. O'Toole et al. (1997) Pituitary-adrenal cortical axis measures as predictors of sustained remission in major depression. *Biological Psychiatry*, 42, 85-89; Rubin et al. (1996) Adrenal gland volume in major depression: relationship to basal and stimulated pituitary-adrenal cortical axis function. *Biological Psychiatry*, 40, 89-97.
3. Beck-Friis et al. (1985) Melatonin, cortisol, and ACTH in patients with major depressive disorder and healthy humans with special reference to the outcome of the dexamethasone suppression test. *Psychoneuroendocrinology*, 10, 173-186.

4. Zobel et al. (2000) Effects of the high-affinity corticotropin-releasing hormone receptor 1 antagonist R121219 in major depression: the first 20 patients treated. *Journal of Psychiatric Research*, 34, 171-181.

Chapter 5 – The Broader Picture: Stress Influences the Monoamine Systems

1. The scenario is thought to work as follows: increased levels of serotonin resulting from the injection lead to a decrease in glucocorticoid concentrations, possibly mediated by an inhibition of the HPA axis (by way of the hippocampus - Figure 5-2). This decrease in circulating glucocorticoid levels results in an eventual increase in glucocorticoid receptors in order to soak up what little glucocorticoid is available. The reverse has also been shown. If you reduce the amount of serotonin being pumped out of the raphe nuclei, by damaging the cell's ability to synthesize or release the transmitter, the most commonly observed result is an increase in glucocoroticoid circulation and an eventual down-regulation of glucocorticoid receptors in various limbic structures.
2. Lopez et al. (1999) Role of biological and psychological factors in early development and their impact on adult life. *Biological Psychiatry*, 46, 1461-1471.
3. In addition to the above-mentioned effects of glucocorticoids, CRF also has a direct effect on serotonin release. Manipulations that increase CRF levels in the central nervous system result in a reduction in the release of serotonin to stressful stimuli. Furthermore, direct infusion of CRF into the raphe nuclei disrupts serotonin release from these cells, indicating the HPA axis and the serotonin transmitter system interact at multiple levels.

Chapter 6 – Birth and Death of a Brain Cell

1. See the following papers for excellent reviews on the subject: McEwen, BS (2000) Effects of adverse experiences for brain structure and function. *Biological Psychiatry*, 48, 721-731;

Sapolsky, RM (2000) The possibility of neurotoxicity in the hippocampus in major depression: a primer on neuron death. *Biological Psychiatry*, 48, 755-765.
2. Gould et al. (2000) Regulation of hippocampal neurogenesis in adulthood. *Biological Psychiatry*, 48, 715-720; Cameron, HA, Gould, E (1994) Adult neurogenesis is regulated by adrenal steroids in the dentate gyrus. *Neuroscience*, 61, 203-209.
3. Nemeroff, CB (1999) The preeminent role of early untoward experience on vulnerability to major psychiatric disorders: the nature-nurture controversy revisited and soon to be resolved. *Molecular Psychiatry*, 4, 106-108.

Chapter 7 - Monoamine Theories of Mood: Revisited & Revised

1. Laboratory experiments have demonstrated that serotonin-containing cells in the raphe nuclei release less transmitter in the presence of norepinephrine (binding to alpha 1 receptors). That is, either norepinephrine or serotonin binding to receptors on serotonin-containing cells will shunt the release of transmitter. However, it should be noted that serotonin-containing cells also have alpha 2 receptors for norepinephrine, which when activated *promote* the release of serotonin.

Chapter 8 – Substance P and the Neurokinins

1. Kramer et al. (1998) Distinct mechanisms for antidepressant activity by blockade of central substance P receptors. *Science*, 281, 1640-1645.
2. Kramer et al. (2000) The therapeutic potential of substance P (NK-1 receptor) antagonists (SPA). *Neuropsychopharmacology*, 23, S22.

Chapter 9 – The Immune System and Mood

1. Zorzenon et al. (1996) Major depression, viral reactivation and

immune system. *European Psychiatry*, 11, 4-32.
2. Seidel et al. (1996) Major depressive disorder is associated with elevated monocyte counts. *Acta Psychiatrica Scandinavia*, 94, 198-204.
3. Sluzewska et al. (1995) Interleukin-6 serum levels in depressed patients before and after treatment with flouxetine. *Annals of the New York Academy of Sciences*, 762, 474-477.
4. Rabkin, JG, Harrison, WM (1990) Effect of imipramine on depression and immune status in a sample of men with HIV infection. *American Journal of Psychiatry*, 147, 495-497.

Chapter 10 - The Immune Response, Stress, and Depression: Is there a Final Common Pathway?

1. On a related note, peripheral cytokines are critical regulators of bone metabolism and likely play a role in cardiovascular disease. Thus activation of peripheral cytokines may contribute to the increased rate of heart disease and decreased bone mineral density often observed to co-occur with affective disorders.

Chapter 11 – The Neuroanatomy of Mood & Melancholy

1. Mayberg et al (2000) Regional metabolic effects of flouxetine in major depression: serial changes and relationship to clinical response. *Biological Psychiatry*, 48, 830-843.
2. Brody et al. (2001) Regional brain metabolic changes in patients with major depression treated with either paroxetine or interpersonal therapy. Archives of General Psychiatry, 58, 631-640; Martin et al (2001) Brain blood flow changes in depressed patients treated with interpersonal psychotherapy or venlafaxine hydrochloride. Archives of General Psychiatry, 48, 641-648.
3. For an excellent review, see Drevets, WC (2000) Neuroimagingstudies of mood disorders. *Biological Psychiatry*, 48, 813-829.

Chapter 13 – Biological Rhythms and Mood

1. Riemann et al. (2001) Sleep and depression – results from psychobiological studies: an overview. *Biological Psychology*, 57, 67-103.
2. King, DP, Takahashi, JS (2000) Molecular genetics of circadian rhythms in mammals. *Annual Review of Neuroscience*, 23, 713-42.

Chapter 14 - New Vistas in our Understanding of Mood and its Disorders

1. Duman et al. (1997) A molecular and cellular theory of depression. *Archives of General Psychiatry*, 54, 597-606.
2. Nibuya et al. (1996) Chronic antidepressant administration increases the expression of camp response element-binding protein (CREB) in rat hippocampus. *Journal of Neuroscience*, 16, 2365-2372.

About the Author

Gene V. Wallenstein, Ph.D. was educated at Harvard University and Florida Atlantic University. He has received major research grants from the National Institutes of Health and the National Science Foundation, and has authored over fifty scholarly articles, abstracts, and book chapters. He is currently an adjunct professor at the University of California, San Diego, and the Director of the Cognitive Neurobiology Institute.

Acknowledgments

I have been blessed by the erudition of so many gifted scientists, teachers, clinicians, and friends who have shaped my thinking over the years, without whom this book would not have been possible. Among those whose influence has been felt the most include: Howard Eichenbaum, Dean Hamer, Michael Hasselmo, Erik Kandel, J. A. Scott Kelso, Raymond Kesner, Melvin Konner, Michael Meaney, Sherri Mizumori, Melissa Monroe, Charles Nemeroff, Jaak Panksepp, Paul Plotsky, Edmund Rolls, Robert Sapolsky, E. O. WIlson, and the late Stephen Jay Gould. I also wish to thank Stephanie Dalton Cowan for her beautiful artwork that graces the cover, and my editor at Commonwealth, Nigel Goodwin for his patience with my creative use of commas and the like.

Index

A

B

C

D

E

F

G

H

I

L

M

N

P

R

S

T

V